SpringerBriefs in History of Science and Technology

More information about this series at http://www.springer.com/series/10085

Angelo Baracca · Rosella Franconi

Subalternity *vs.* Hegemony, Cuba's Outstanding Achievements in Science and Biotechnology, 1959–2014

 Springer

Angelo Baracca
Department of Physics and Astronomy
University of Florence
Sesto Fiorentino, Florence
Italy

Rosella Franconi
Department for Sustainability
ENEA, Italian National Agency for New
  Technologies, Energy and Sustainable
  Economic Development, Casaccia
  Research Centre
Rome
Italy

ISSN 2211-4564　　　　　　ISSN 2211-4572　(electronic)
SpringerBriefs in History of Science and Technology
ISBN 978-3-319-40608-4　　　ISBN 978-3-319-40609-1　(eBook)
DOI 10.1007/978-3-319-40609-1

Library of Congress Control Number: 2016943064

Printed on acid-free paper

This Springer imprint is published by Springer Nature
The registered company is Springer International Publishing AG Switzerland

Acknowledgements

AB is grateful to the Department of Physics of the University of Florence for financial support. AB is also indebted to the Director of the Max Planck Institute for the History of Science in Berlin, Prof. Jürgen Renn, for his interest, hospitality and support during the beginning of this research.

All the information on the development of physics in Cuba originates, and is quoted, from the previous comprehensive work published in the volume A. Baracca, J. Renn and H. Wendt (eds.), *The History of Physics in Cuba*, Berlin, Springer, 2014. AB is deeply indebted towards all the Cuban colleagues who collaborated in that research.

AB is also indebted to Edoardo Magnone for his encouragement in the early steps of this research, and further information on science in South Korea. We are indebted to Paolo Amati for his witness and help that stimulated our research, still in progress, on the role of Italian geneticists on the development of Cuban biotechnology. We are grateful to Marina and Luciano Terrenato, University of Rome "La Sapienza", for proving us information and original documents about the 1971 "Summer school" in Genetics, and to the haematologist Gisela Martinez for providing information about Bruno Colombo.

RF is grateful to her Cuban colleagues, in particular the scientists of the International Centre of Genetic Engineering and Biotechnology (CIGB).

Contents

1   **Introduction. Cuba's Exceptional Scientific Development**   1
     1.1   An Epochal Thaw   1
     1.2   The Gramscian Concept of Hegemony Applied to the Case
          of Cuba   2
     1.3   Cuba's Leap Forward in the Sciences.   4
     1.4   An Unconventional, Open-Minded Attitude.   5
     1.5   Ends Before Means   6
     1.6   International Recognition of Cuba's Achievements
          in the Field of Biotechnology   6
     1.7   What Will the Future Hold?   8
     References   9

2   **Meeting Subalternity, A Constant Challenge in Cuban History**   11
     2.1   Cultural Emancipation as a Condition for Full Independence....   12
     2.2   A Coherent Intellectual Path   12
     2.3   Early Cuban Advances in Medicine   14
     2.4   An Aspect of Subalternity: Early Introduction of Advanced
          Technologies *Versus* a Delay in Basic Sciences.   15
     2.5   The Forging of a National Identity, the Ideas of "Cubanity"   17
     2.6   The Frustration of US Occupation   18
     2.7   Social and Cultural Ferments Under US Rule   20
     2.8   The Weight of Subalternity. Contrasts in Pre-revolutionary
          Cuba   21
     2.9   *Granma* Disembarks the Revolutionary Leaders.   22
     References   23

3   **Addressing the Challenge of Scientific Development:**
  The First Steep Steps of a Long Path   25
     3.1   A Future of Men and Women of Science   26
     3.2   Free Education   27
     3.3   University Reform, Fostering Scientific Research.   28
     3.4   Early Student-Led Updating of the Teaching of Physics   29

3.5   Students to the Soviet Union.   30
3.6   Fostering Research in Physics as a Strategic Choice,
        Taking Advantage of All Sources of Local
        and Foreign Support.   30
3.7   Leaps Forward in Reaction to Ominous Threats.   32
3.8   Another Strategic Cornerstone: Promoting Medicine
        and Health Care   33
3.9   The Cuban Academy of Sciences.   35
3.10   The National Centre for Scientific Research   36
References   36

4   Reaching a Critical Mass and Laying the Foundations
    of an Advanced Scientific System   39
    4.1   Rapid Achievements in Science.   39
    4.2   Participated and Socially-Oriented Discussion of Scientific
            Choices   41
    4.3   A Network of Specialized Technical Scientific Institutions   43
    4.4   Summer Schools and Achievements in Microelectronics.   45
    4.5   Overall Progress in Physics.   46
    4.6   The Decisive Italian Support to the Development of Modern
            Biology in Cuba   47
    4.7   Growing Institutional Planning of the Cuban
            Scientific System   51
    References   52

5   The Decisive Leap in the 1980s: The Attainment of Cuba's
    Scientific Autonomy.   55
    5.1   New Planning of Scientific Development, with the Goal
            of Reaching Autonomy.   56
    5.2   Fostering Electronics, and "Improvising" Superconductivity   57
    5.3   The Project of a Nuclear Power Plant: Nuclear Physics
            as the Backbone of Cuban Scientific System.   59
    5.4   Redirecting Scientific Development   61
    5.5   The Growing Strategic Role of Biotechnology for Achieving
            Autonomy.   61
    5.6   Entering Modern Biotechnology from Its Beginnings: Obtaining
            Interferon for the Country's Own Needs.   63
    5.7   The Leap Towards Genetic Engineering   65
    5.8   Ends Above Means: Differentiating from Mainstream Western
            Biotechnology.   66
    5.9   The First Great Achievements and Further Implications
            of a Need-Driven Approach   68
    5.10   A Sound Network of International Relations   70
    5.11   An Integrated Biomedical Network   71
    References   72

6 Decisive Results ... and New Challenges 75
   6.1 A "Disaster Proof" Scientific System 76
   6.2 Meeting a New Challenge. 77
   6.3 Further Impulse to the Cuban Scientific System. 79
   6.4 More Challenging Choices 81
   6.5 More Recent Achievements. 82
   6.6 Further Cuban Distinctive Features: South–South
        Cooperation, Medical Diplomacy...................... 85
   6.7 Cuba's Remarkable and Enduring Achievements 88
   References 90

7 Comparative Considerations and Conclusions 93
   7.1 The Intriguing Issue of Cuba's Scientific Achievement:
        Knowledge-Based Economy and State High Technology
        Company 93
   7.2 Peculiar Features of Cuban Biotechnology Industry 96
   7.3 Something Worth Thinking Seriously About: A Comparison
        with Other Experiences 98
   7.4 Conclusions 101
   References 102

Abbreviations and Acronyms

ACC	Cuban Academy of Science
BIOCEN	National Centre of Bioproduction
CBFM	Centre for Biophysics and Medical Physics
CEAC	Cuban Commission for Atomic Energy
CELAC	Community of Latin American and Caribbean States
CEADEN	Centre for Studies Applied to Nuclear Development
CENPLAB	National Centre for Production of Laboratory Animals
CIB	Centre for Biological Research
CIGB	Centre for Genetic Engineering and Biotechnology
CIM	Centre for Molecular Immunology
CLAF	Latin American Centre for Physics
CNC	Cuban Centre of Neurosciences
CNIC	National Centre for Scientific Research (in some publications, CENIC)
COMECON (or CMEA)	Community for Mutual Economic Assistance
CQF	Chemical-Pharmaceutical Centre
CUJAE	Ciudad Universitaria (Politechnical University) "José Antonio Echeverría"
EGF	Epidermal growth factor
ELAM	Latin American School of Medicine
GATT	General Agreement on Tariffs and Trade
IFN	Institute for Nuclear Physics
IMRE	Institute of Materials Science and Technology
ININ	Institute for Nuclear Research
ININTEF	Institute for Fundamental Technical Research
INOR	National Oncology and Radiobiology Institute
INRA	National Institute for Agrarian Reform
IP	Intellectual property
IPK	Institute of Tropical Medicine "Pedro Kourí"
IPVCE	Exact Sciences Vocational Senior High Schools
ISCM-H	Medical Sciences Higher Institutes in Havana

ISCM-VC   Medical Sciences Higher Institutes in Villa Clara
ITM   Military Technical Institute
MES   Ministry of Higher Education
MIT   Massachusetts Institute of Technology
SEAN   Executive Secretariat for Nuclear Affairs
SLAFES   Latin American Symposium on Solid State Physics
SUMA   Ultra Micro Analytic System
TRIPS   Agreement on Trade-Related Aspects of Intellectual
                  Property Rights
UNDP   United Nations Development Program
UNESCO   United Nations Educational, Scientific and Cultural
                  Organization
UNIDO   United Nations Industrial Development Organization

Chapter 1
Introduction. Cuba's Exceptional Scientific Development

Abstract The current thaw between the United States and Cuba is a major focus of worldwide attention and discussion. Among all the aspects of this interest towards Cuba, there is one that will presumably receive scant attention, but that has for many years been a topic of great interest within specialized scientific milieus: i.e., the fact that this small island, despite scarce resources and the disadvantages arising from the longest embargo in modern history, has almost incredibly reached an exceptional scientific level, in particular in the field of biotechnology, a typically capital-intensive and American-controlled field. Indeed, the goal of developing an advanced scientific and educational system was a specific priority of the revolutionary government from the outset in 1959. The declared aim of this program was to overcome the condition of subalternity that usually dooms developing countries to perpetual dependence. This ambitious project has been achieved through a highly original approach, an open-minded attitude that has put the needs of the population and of the nation before every other consideration.

Keywords Cuba · Gramsci's concept of hegemony · Scientific development · Development *versus* underdevelopment · Cuban biotechnology · South-South cooperation · Cuban health system · Hegemony *versus* subalternity · Embargo

1.1 An Epochal Thaw

Cuba has once again burst into the limelight of the international stage after President Barack Obama's unexpected announcement at the end of 2014 that he intended to remove the last ghost of the Cold War by re-establishing political relations with Cuba after more than half a century of an anachronistic embargo and of countless attempts to stifle the Cuban economy. At the time of our writing (December 2015) one year has passed from that announcement and the embargo is still on, while Cuba is once more the focus of intense (even if presumably far from unbiased) political and economic interests and of active initiatives from all around the world.

© The Author(s) 2016

A. Baracca and R. Franconi, *Subalternity vs. Hegemony, Cuba's Outstanding Achievements in Science and Biotechnology, 1959–2014*, SpringerBriefs in History of Science and Technology, DOI 10.1007/978-3-319-40609-1_1

1

Apart from political appraisals and predictions, the current situation offers a good opportunity to speak about Cuba from a new perspective in order to assess the post-revolutionary Cuban experience. In particular, there are certain aspects of this experience that are unlikely to attract general attention but that unquestionably represent enduring achievements. What is more, the Cuban revolution has reached these achievements following rather uncommon paths. These aspects can therefore be discussed without going into the tickling question of an assessment of the Cuban political and economic "regime", about which different opinions are certainly legitimate, as are certain concerns (but of what country, ultimately, is this not true?).

The specific aspect we wish to discuss here is the following:

i. from the earliest moments after the victory of the revolution (1959), despite highly adverse conditions, Cuba has made an enormous effort to definitively overcome its condition of *subalternity* and acquire substantial autonomy;
ii. this purpose was actually attained in a surprisingly short time, thanks to the resolute choice of developing an advanced scientific system, a project that might have seemed unrealistic considering the country's initial conditions, but has instead been completely successful;
iii. not less remarkable in this process is that every choice has constantly been driven by the basic needs of the population and of the country's social economic development.

Cuba's achievement of advanced scientific development is an exceptional case among underdeveloped countries. It is even more striking if one takes into account the country's peculiar features. We are speaking, in fact, about a small Caribbean island that gained independence (though admittedly conditional independence) just over a century ago. It covers less than one-thousandth of the earth's surface, houses barely 1.5 parts per thousand of the world population and has roughly only one thousandth of the world GDP. Yet Cuba has influenced international assets and events in a measure disproportionate to its apparent "insignificance". One example for all—the recent visit of Pope Francis (September 2015) was the third time a Pope has visited Cuba, a world record. It should be added that the worldwide influence of this strip of land is not restricted to political events. Rather it extends to various cultural fields, even if they are sometimes ignored: we need only remember here the originality and worldwide influence of Cuban African-American music and rhythm (as was brought to general attention by Wim Wenders' movie "Buena Vista Social Club").

1.2 The Gramscian Concept of Hegemony Applied to the Case of Cuba

Before entering into our analysis about how Cuba was able to develop this innovative capacity, we must dedicate some words to our adoption of the concepts of *subalternity* and *hegemony* in order to focus our discussion. Marx had discussed the economic conditions for the proletarian revolution. As is widely known, the concept of cultural hegemony was created by the Italian Marxist politician and philosopher Antonio Gramsci (1891–1937), who insisted on the need for the proletariat to cut loose from the cultural hegemony of the dominant classes and to achieve its own cultural hegemony (Storey 1994). In fact, according to Gramsci, the dominant classes succeed in imposing a consensus about their own definition of reality, their world view, so that it is accepted by other classes as "common sense". In Gramsci's words:

> the supremacy of a social group manifests itself in two ways, as 'domination' and as 'intellectual and moral leadership'", and "the 'normal' exercise of hegemony on the now classical terrain of the parliamentary regime is characterized by the combination of force and consent, which balance each other reciprocally, without force predominating excessively over consent. (Gramsci 1971, p. 215; original publ. Gramsci 1948–1951, p. 70)

The revolution, therefore, must be accompanied by the conquest of cultural hegemony, securing the class of traditional intellectuals to the proletariat and making them its own political leaders. After the victory of the revolution political leaders will have to assure the continuation of the cultural hegemony of the proletariat. While in the present study on the development of science in Cuba we will not enter into theoretical considerations, the concept of *hegemony vs. subalternity* will prove particularly useful in discussing Cuba's choices.

The challenge of getting beyond a condition of *subalternity* has been a crucial one for all underdeveloped countries. Each country has followed its own path, but not many, even those much larger than Cuba, have reached real autonomy from the leading world powers (meaning basically, at present, the United States, which in the post-World War II years has replaced the political, economic and technological dominance of the former colonial countries, i.e., the UK, France and Germany). What is more, the countries that have achieved considerable (though not full, and in different degrees) level of autonomy are mainly large, highly populated ones, such as China, India and Brazil. For smaller countries, the challenge remains substantially unmet.

Moreover, for most developing countries filling the gap is virtually impossible because of the insurmountable difficulty of keeping up with the speed of innovation in the developed countries. Japan and South Korea represent relevant exceptions, in that they have succeeded in reaching impressive and rapid development and even in creating their own industrial and technological empires. However, it must be added that in order to boost their development both of these countries have relied almost totally on full adhesion to the economic and technological model, as well as to the values, of the United States, as well as on its unstinting support (given basically for geopolitical reasons). Similarly, the development of the eastern European countries benefited from their adhesion to the Soviet-led Council for Mutual Economic Assistance (COMECON, or CMEA). However, in both cases this choice subjected these countries to the ups and downs of the leading power of reference, of the global economy, and of their respective target markets. So, for instance, South Korea was hard hit by the unexpected crisis of the "Asian Tigers" in 1997, and Japanese economic power had also been declining in recent years. Not to mention the European socialist countries which, after the dissolution of the Soviet Union and the Warsaw Pact, suffered a profound crisis that forced them to radically change their economic and industrial structures. After all, when all is said and done, this is just what happened to the older European powers when the United States imposed its supremacy after World War 2.

1.3   Cuba's Leap Forward in the Sciences

Cuba is one "small" but relevant exception to what we have said above. Since the victory of the revolution in 1959, though starting out from a very difficult situation, this country has succeeded in overcoming its condition of *subalternity* in a distinct, largely autonomous and original way, and in the relatively short span of a few decades. This success is the focus of this book.

With regard to the difficult initial conditions in the field of the sciences, as Clark Arxer said,

> A report by the ad hoc Truslow Commission of the International Bank for Reconstruction and Development, which had travelled to Cuba to study the provision of loans, stated unequivocally in 1950 that 'in the field of applied research and labs, there was no development at all in Cuba'. (Sáenz and García-Capote 1989; Clark Arxer 2010).

Yet in only a few decades Cuba has reached levels of international excellence and a condition of scientific autonomy in several domains, in particular, but not only, in the bio-medical field. As the same 2010 UNESCO Science Report emphasizes,

> By the dawn of the 21st century, Cuba was perceived as being a proficient country in terms of scientific capacity, despite having experienced more than four decades of a trade embargo and restrictions on scientific exchanges imposed by successive US administrations (Jorge-Pastrana and Clegg 2008). In a study commissioned by the World Bank in 2001, Wagner et al. of RAND, an S&T think tank in the U.S. classified nations into four categories according to their scientific prowess: developed, proficient, developing and lagging. In Latin America and the Caribbean, only Brazil and Cuba qualified as 'proficient'. (Clark Arxer 2010).

Today, the percentage of university graduates and physicians in the Cuban population of just over eleven million and the overall level of their scientific training rivals that of many highly developed nations and has no equals among other underdeveloped countries (Hoffmann 2004, pp. 166–168). In 1959 there were only a few dozen physicists in the whole country. And immediately after the revolution, by the middle of 1960, more than 20 % of professionals and technicians and almost one half of the slightly over 6000 Cuban doctors had left the country (Martínez Pérez 2006, p. 72). Yet today Cuba has wiped out all the third world diseases and boasts a first-world health profile. There is a mid-level technician for every eight workers, a university graduate for every fifteen workers and 590 physicians for every 100,000 inhabitants; there are over 160 centres of scientific research, and 1050 engineers and scientists for every million inhabitants (CEDISAC 1998). The Western Havana Bio-Cluster employed 12,000 workers and more than 7000 scientists and engineers in 2006 (Lage 2006). BioCubaFarma, the Biotechnology and Pharmaceutical Industry Group created in 2012 in order to promote businesses related to medical services, currently consist of 32 entities, 78 production facilities and employs almost 22,000 workers.[1]

1.4 An Unconventional, Open-Minded Attitude

No less interesting is *how* Cuba has reached such results. At first sight, one could think that the rapid scientific development of the country and its attainment of a First-World scientific profile was completely due to the unconditional support provided for almost three decades by the Soviet Union. Yet, though the importance of Soviet support could hardly be underestimated, it was certainly not the only factor at work, and in some sectors not even the main one.

As a matter of fact, the shaping of the Cuban scientific system was a far more original, complex and multifaceted process. The Cuban scientific community was open to, and took advantage of, diverse schools of research and sources of support and collaboration besides the Soviet one, in particular "western" scientists and nations. Cuban scientists were able to integrate these different contributions into an original process of constructing a sound, well-structured, integrated and advanced scientific system. There was even one case, the field of biological sciences, in which Soviet science could be of no help at all since, for eminently ideological reasons, it long refused modern developments in genetics and molecular biology. Still, Cuba has reached a leading position in the typically American-dominated and capital-intensive field of biotechnology by resorting to support from western scientists and institutions, integrated with typical Cuban resourcefulness and originality.

The success and solidity of the resulting scientific structure became evident with the disintegration of the Soviet Union and the socialist market. Contrary to most predictions, not only the Cuban scientific system but the country's overall economic and political structure successfully resisted this tremendous shock (thus representing the only exception in the entire socialist block). Once more, quotations from Gramsci are useful to interpret this outcome:

> Every social group... creates together with itself, organically, one or more strata of intellectuals which give it homogeneity and an awareness of its own function not only in the economic but also in the social and political fields. (Gramsci 1971, p. 217; original publ. Gramsci, 1948–1951, p. 72).

---

[1]http://oncubamagazine.com/economy-business/biocubafarma-unite-and-conquer/. Last access March 15, 2016.

and,

> One of the most important characteristics of any group that is developing toward domi-
> nance is its struggle to assimilate and to conquer 'ideologically' the traditional intellectuals.
> (Gramsci 1971, p. 218; original publ. Gramsci, 1948–1951, p. 73).

1.5 Ends Before Means

In this process of development priority was constantly given to the most urgent
social problems of the population and to the economic needs of the country.
Examples of the correspondence between science and public needs are the Cuban
healthcare system, with the special attention it has given to "third world" diseases,
and the close cooperation between meteorological institutes and civil protection.
This coordination has allowed Cuba to avoid or reduce to a minimum the victims of
natural catastrophes such as tropical hurricanes, which usually claim a far higher
number of victims in neighbouring Caribbean countries (as well as in the United
States). As one scholar has remarked,

> Cuban socialist science has differed from Soviet socialist science. Within Soviet science,
> the means of scientific research were privileged over the ends ... In Cuba, by contrast, the
> ends were valued over the means ... The state's influence on science here would take a
> different path. ... while innovation was still far from a priority.... improvisation was
> already a valued ethic among the growing ranks of scientists and technicians. (Reid-Henry
> 2010, p. 27).

Cuba's approach has also been considered an exceptional example of
"South-South cooperation". A significant example, in health biotechnology, is
given by the entrepreneurial cooperation with Brazil that led to the joint production
of a meningitis vaccine for Africa (Sáenz et al. 2010; Cortes et al. 2012). In addition
to its medical missions in Latin America, Africa and in countries hit by natural
disasters, Cuba brings thousands of students from one hundred countries to study
medicine at the Latin American School of Medicine (ELAM) at no cost (Fitz 2011).
After six years of training at ELAM, graduating doctors take the model of primary
and preventive health care back to the distressed communities that need it most,
helping to turn the "brain drain" into a "brain gain."

1.6 International Recognition of Cuba's Achievements
    in the Field of Biotechnology

Before proceeding further, some clarifications seem necessary. In the first place, the
tone we are using to describe Cuban scientific achievements may sound overly
eulogistic, and our analysis one-sided. We should like to clarify that the words we

use in what follows do not imply value judgements, least of all about the Cuban political and economic system. Our purpose is, in fact, more limited: i.e., to discuss, on the basis of verifiable historical facts and data, the achievements of Cuban science and the particular features of the approach used and choices made in its development. However, even a simple evaluation of Cuba's scientific achievements cannot fail to take into consideration the small size of the country, its limited resources and the extremely difficult conditions under which it has been forced to operate.

As we shall see in detail in the following analysis, Cuba's scientific results in the field of biotechnology and the originality of its approach are acknowledged by authoritative and independent scientific sources like *Science, Nature* and others (Kaiser 1998; Thorsteinsdóttir et al. 2004; Buckley et al. 2006; López Mola et al. 2007; Evenson 2007; Editorial 2009; Starr 2012; Fink et al. 2014). Specific assessments of the development of Cuban science in various fields have previously been presented: for *biotechnology* in López et al. 2006, 2007; Cárdenas 2009; Reid-Henry 2010; and for the development of *physics* in Baracca et al. 2014b. We will frequently refer to these works in our analysis, while avoiding excessive detail. However, we wish to stress that while previous studies have considered separately the development of either physics or biotechnology in Cuba, in the present analysis we will refer, at times comparatively, to the development of both sectors and in particular to the underlying reasons for Cuba's commitment to biotechnologies. It is our hope that this integrated discussion will shed more light on the ultimate goals pursued by the Cuban leadership and scientific community in promoting advanced technical scientific development, and on the results they have reached. For instance, in the case of biotechnology previous studies have tried to weigh Cuban domestic needs against commercial mechanisms. This point of view can be greatly enlarged, and perhaps even changed, by taking into account the initial project, which preceded the commitment to modernize the biological sciences, of developing a modern physics sector as a strategic choice in order to provide a sound foundation for other scientific fields.

Moreover, the present study fills an existing gap regarding the process of training and updating of Cuban scientists during the 1960s and the early 1970s, which was a necessary precondition for subsequent scientific development. This process has already been investigated for physicists in Baracca et al. (2014a, b). In this study the training of biologists is clarified, thanks to interviews with the Italian biologists (in particular, Paolo Amati) who in the early 1970s played a crucial role by promoting intensive six-month courses, coordinated with Cuban authorities, for the most promising Cuban students, some of whom were subsequently given the possibility to specialize in Italian institutions. Some of these later went on to become leading figures in the future Cuban biotechnological complex.

1.7   What Will the Future Hold?

Nobody can say what the future will hold. What is certain is that nothing will be as it was before. Cuba is not new to or unprepared for epochal upheavals. The shock that followed the (unexpected) collapse of the Soviet Union at the turn of 1990s strained the conquests of the revolution to the limit. Since then the country has sailed the high seas, with no friends in important places. The current situation is no less uncertain, for it presents both opportunities and dangers. The global asset has radically changed since 1990, and at the present moment it faces even greater uncertainties. The real purpose of the current American opening towards Cuba is not clear, and it is probably far from unequivocal. Obama is at the end of his presidency. Who will succeed him? His thaw has strong opponents. The process of liberalization of the world economy is undergoing unprecedented acceleration with the projects of the Trans-Pacific (TIP) and Trans-Atlantic (TTIP) treaties. What new challenges will Cuba have to face in the future?

International power relations will presumably undergo deep changes as well. In recent years Cuba has heavily relied on alliances with, and support from, several Latin American countries, at the same time as the pressures exerted by the United States have been weakened. But at the present moment the wind of renewal in the sub-Continent seems to be declining, considering the recent political elections in Argentina and Venezuela and the increasing difficulties being faced by the President of Brazil, Dilma Roussef. In the future Cuba may risk losing the support it presently has from these countries.

Moreover, the historical Cuban leadership has arrived at the end of the road. Its future replacement is fraught with uncertainty and could have surprises in store. Indeed, the present transition probably represents one of the most critical crossroads in Cuban history.

In defiance of all this, Cuba is playing an increasing international role, not only as a door to America but also as an interface between two worlds, contributing to solve conflicts or settle controversies. For instance, from 2012 it harbours the peace talks (started in Oslo) between the Colombian government and the FARC rebel movement, broking the information and diplomatic blockade controlled by the US. Very recently (February 12, 2016), Pope Francis, after his official visit to Cuba in 2015, and the Patriarch of Russian Orthodox Church, Kirill (who already in 2008, when he still was Patriarch, met Fidel Castro) have chosen Havana for a historic first meeting for the heads of the two Churches after a millennium-long rift.

For what concerns the main subject of this book, Cuban biotechnology, it seems to have reached a crossroads as well: the input of foreign capital seems unavoidable in order to meet competition that promises to be increasingly fierce. In this direction, in 2014, the government created the Financial Fund for Science and Innovation (FONCI) to enhance the socio-economic and environmental impact of science by boosting business innovation. This is a major breakthrough for Cuba, considering that, up to now, the bulk of R&D funding has come from the public purse (UNESCO Science report, November 2015). We ourselves have sensed a degree of uneasiness among some members of the Cuban biotechnology community in the face of looming changes. It is likely that in this field, too, nothing will be as before. This is why we have deliberately limited our present reconstruction of the evolution of the Cuban scientific structure to the period going from 1959 to 2014.

At present everything is fluid. More than one year after Obama's political opening, the greatest problem for Cuba remains the removal of the anachronistic embargo. But in fact nothing has changed in this regard, since the opposition within the US seems insurmountable. Yet, in the latest round of the annual UN vote on the embargo to Cuba on 28 October 2015, the United States was left holding the short end of the stick, voting with the only company of Israel.

References

Baracca A, Fajer Avila V, Rodríguez Castellanos C (2014a) A comprehensive study of the development of physics in Cuba from 1959 (Baracca A, Renn J, Wendt H, eds) 2014:115–234

Baracca A, Renn J, Wendt H (eds) (2014b) The history of physics in Cuba. Springer, Berlin

Buckley J, Gatica J, Tang M, Thorsteinsdóttir H, Gupta A, Louët S, Shin Min-Chol, Wilson M (2006) Off the beaten path. Nat Biotechnol 24:309–315

CEDISAC (1998) Todo de Cuba. Madrid: CEDISAC, Prensa Latina. [Multimedia encyclopedia]

Clark Arxer I (2010) Cuba. In: UNESCO science report 2010, Chapter 6, pp 123–1331. http://www.unesco.org/new/en/natural-sciences/science-technology/prospective-studies/unesco-science-report/unesco-science-report-2010. Last access 15 July 2014

Cortes Mde L, Cardoso D, Fitzgerald J, DiFabio JL. 2012. Public vaccine manufacturing capacity in the Latin American and Caribbean region: current status and perspectives. Biologicals Jan; 40(1):3–14. doi:10.1016/j.biologicals.2011.09.013

Cárdenas A (2009) The Cuban biotechnology industry: innovation and universal health care. https://www.open.ac.uk/ikd/sites/www.open.ac.uk.ikd/files/files/events/innovation-and-inequality/andres-cardenas_paper.pdf. Last access 15 March 2016

Editorial (2009) Cuba's biotech boom. The United States would do well to end restrictions on collaborations with the island nation's scientists. Nature. 457 (January): 8

Evenson D (2007) Cuba's biotechnology revolution. MEDICC Review 9(1):8–10

Fink GR, Leshner AI, Turekian VC (2014) Science diplomacy with Cuba. Science 344 (6188):1065

Fitz D (2011) The Latin American school of medicine today: ELAM. Monthly Review 62(10). http://www.nnoc.info/latin-american-school-of-medicine/. Last access 14 March 2016

Gramsci A (1971) Selections from the prison notebooks. Hoare Q, Nowell Smith G (eds) International Publishers, New York; original publication, Quaderni del Carcere, Il Risorgimento, F. Platone ed., Torino, 1948–1951

Hoffmann B (2004) The politics of the internet in third world development. Challenges in contrasting regimes with case studies of Costa Rica and Cuba. New York, Routledge

Jorge-Pastrana S, Clegg M (2008) US-Cuban scientific relations. Science 322, 17 October, p. 345

Kaiser J (1998) Cuba's billion-dollar biotech gamble. Science 282(5394):1626–1628

Lage A (2006) Socialism and the knowledge economy: cuban biotechnology. Monthly review 58(7). http://monthlyreview.org/2006/12/01/socialism-and-the-knowledge-economy-cuban-biotechnology/. Last access 14 March 2016

López Mola E, Silva R, Acevedo B, Buxadó JA, Aguilera A, Herrera L (2006) Biotechnology in Cuba: 20 years of scientific, social and economic progress. J Commercial Biotechnol 13:1–11

López Mola E, Silva R, Acevedo B, Buxadó JA, Aguilera A, Herrera L (2007) Taking stock of Cuban biotech. Nature Biotechnol 25 (11, November): 1215–1216

Martínez Pérez L (2006) Los Hijos de Saturno. Intelectuales y Revolución en Cuba. Facultad Latinoamericana de Ciencias Sociales, Sede México, D.F.

Reid-Henry S (2010) The Cuban cure: reason and resistance in global science. University of Chicago Press, Chicago

Starr D (2012) The Cuban biotech revolution. http://www.wired.com/wired/archive/12.12/cuba_pr.html. Last access 15 March 2016

Storey J (ed) (1994) Cultural theory and popular culture: a reader. Harvester Wheatsheaf, NY

Sáenz T, García-Capote E (1989) Ciencia y Tecnología en Cuba. Editorial in Ciencias Sociales, Havana

Sáenz TW, Thorsteinsdóttir H, de Souza MC (2010) Cuba and Brazil: an important example of South-South collaboration in health biotechnology. MEDICC Rev. Jul;12(3):32–35

Thorsteinsdóttir H, Sáenz TV, Quach U, Daar AS, Singer PA (2004) Cuba. Innovation through synergy. Nat Biotechnol 222 (Supplement) December: 19–24

Chapter 2
Meeting Subalternity, A Constant Challenge in Cuban History

Nothing is more similar to the myth of the bird phoenix than the social and political history of Cuba during the past century. From 1898 to our days, the country has dealt with a rebirth approximately every 30 years: from the North-American occupation of the island as a solution to the war of independence, to the revolution of 1930, from the latter to the Revolution of 1959, and from that to the economic crisis and the consequent reconsideration of the social model of the country caused by the disappearance of real socialism of Eastern Europe, begun in 1989.

[Martínez Pérez 2006, 9]

Abstract The need to overcome the condition of subalternity—first from the colonial dominance of Spain, and then from the economic and political hegemony of the United States—in order to gain true independence, underlay the thought and practice of Cuban freedom-fighters throughout the 19th and 20th centuries. Exponents such as Félix Varela, José Martí, Enrique José Varona, Manuel Gran and Ernesto Guevara were aware that the spread of culture and the development of modern scientific education and research were essential, not only in order to gain political independence but also for the crucial challenge that would follow, i.e., cutting loose from the condition of subalternity. This challenge was closely interwoven with the shaping of a particular national and cultural identity, commonly called *cubanía* (Cubanity), a blend of Spanish and African cultural influences. Under US rule and the bloody dictatorships that characterized the 1930s and 1950s, Cuba underwent a profound social and cultural ferment that was to prepare the country for the great upheaval triggered by the handful of young guerrillas who adventurously disembarked from the boat *Granma* on 2 December 1956.

Keywords Subalternity · Hegemony · Cubanity · Transculturation · Yellow fever · José Martí · Carlos Finlay · Spanish-American war

2.1   Cultural Emancipation as a Condition for Full Independence

Although the exceptional results of Cuban science have been obtained since the victory of the Revolution, one can trace the early roots of awareness of the need to overcome the condition of subalternity to the past history of Cuba, which shows many particular and original features compared to all the other Latin American and Caribbean countries.[1] A first evident fact is that Cuba was the last of these countries to reach independence.[2] After the independence of the thirteen British colonies in 1776, the French Revolution of 1789 and the subsequent independence of Haiti as well as the Napoleonic occupation of Spain led to the independence of Argentina in 1810–1816, Paraguay in 1811, Venezuela in 1811–1819 (in the context of the 'Gran Colombia', which in 1830 was divided into Ecuador, Venezuela and Colombia), Chile in 1818, Peru in 1821 (which Bolivia separated from in 1825) and Mexico in 1821–1823. By contrast, Cuba did not free itself from Spanish colonial dominion until 1898, to pass, after the Spanish-American War, under the hegemony of the new emerging imperial power of the Unites States. Under Spanish rule, the royal power strongly opposed and prevented the development of cultural autonomy and of a modern education system in Cuba. All the more significant it is, then, that the most representative Cuban figures of the 19th and 20th centuries—such as Félix Varela, José Martí, Enrique José Varona, Manuel Gran and Ernesto Guevara[3]— were aware that the diffusion of culture and the development of modern scientific education and research were essential not only in order to get real political independence, but also for the following challenge of cutting loose from the situation of subalternity.

2.2   A Coherent Intellectual Path

At the beginning of the 19th century, the Catholic priest, Félix Varela (1788–1853), who is said to have first taught Cubans to think (Rodríguez 1944), introduced the innovating spirit of the Enlightenment in Cuba, under the enlightened and

---

[1] An extremely useful collection of documents of Cuban authors and short essays on Cuba's history, politics and culture is in: Chomsky et al. (2003).

[2] A personal view of the peculiarities of Cuba's modern history, with special attention to the aspects of its cultural and scientific development, has been discussed by A. Baracca in "The Cuban 'exception': the development of an advanced scientific system in an underdeveloped country", in the volume Baracca et al. (2014, 9–50). This introduction is followed by "A short critical bibliographical guide", by D. Basosi.

[3] Guevara was not actually Cuban. He was born in Argentina, but he played a primary role in the Cuban Revolution and in its further developments, and is usually associated with Cuba.

progressive direction of Bishop Espada,[4] and introduced modern contents of physics as early as 1817 (Torres-Cuevas 1995; Altshuler and Baracca 2014). However, his consciousness was much broader: indeed, when he was elected in 1822 as a representative to the Spanish *Cortes*, he voted in favour of partial autonomy of Cuba from Spain and wrote an influential treatise in favour of the abolition of slavery. As a consequence of these positions, he had to seek refuge in the United States and came to the conclusion that full independence was the only solution. Varela shared the destiny of exile with other intellectuals of this time, like José Maria Heredia (1803–1839), the first great Cuban poet.

José Martí (1853–1895) deserves the credit of having been the first (not only in Cuba, but for the whole of Latin America) to clearly develop full consciousness of the strict connection between culture and power, the indissoluble tie between the attainment of political independence, real democracy and justice without slavery, and emancipation from the condition of subalternity. He not only became the inspirer and leader of the Cuban independence movement, but also was one of the great turn-of-the-century Latin American intellectuals, one of the most influential orators and writers of that period and a forerunner of Modernism in literature. Although Martí never lived to see Cuba free (he was killed on May 19, 1895 in the first battle in which he took part after landing in Cuba to take part to the war with Spain), he is considered the great national hero: his busts and portraits are found everywhere in Cuba. Forced by the colonial regime to live at length in the United States, he could assert: "I have lived in the monster and I know its entrails".[5] Travelling in Mexico, Guatemala and Venezuela, he realized the poor results the popular masses had obtained with independence. Martí perfectly grasped the real contents of US "democracy", and was the first one who understood with great lucidity the roots of US imperialism and the expansionist ambitions that already predominated in US government circles: once the "conquest" of the West was completed, the United States was preparing to expand towards the Antilles and Latin America. This convinced him of the urgency of the liberation of Cuba, in order to prevent this expansion, which would decide the destiny of the Continent. With this aim he launched a heartfelt call to the whole southern Continent in his *Nuestra América* (Our America, 1891), an expression which radical movements have at present taken up again all over the Continent:

---

[4]Juan José Díaz de Espada y Fernández de Landa (1756–1832), who had taken up his diocesan post at the beginning of 1802, was an enlightened person, who waged the struggle against Scholasticism (Figueroa y Miranda 1975).

[5]José Martí, letter to Manuel Mercado, May 18, 1895, http://www.historyofcuba.com/history/marti/mercado.htm. Last access March 16, 2016.

... the pressing need of our America is to show itself as it is, one in spirit and intent ... The scorn of our formidable neighbour who does not know us is our America's greatest danger. And since the day of the visit is near, it is imperative that our neighbour knows us, and soon, so that it will not scorn us.... Once it does know us, it will remove its hands out of respect.[6]

From our point of view, it is important to note that Martí emphasized the importance of education as a crucial factor in the formation of the Cuban nation, independent from Spanish and US educational systems (Quiroz 2006; Strong 2007). Unlike Simón Bolívar, who still relied on the Enlightenment concept of education as an individual form of liberation, Martí was inspired by US-American and British models. He specifically proposed science education, the study of nature, as an instrument for individual autonomy, and the way for promoting social progress, because "to study the forces of nature and learn to control them is the most direct way of solving social problems" (Martí 1953, I, 1076). He thought that Cuba could achieve real independence only when the necessary skills were developed to overcome the economic, political, social and technical underdevelopment inherited from the Spanish colonial regime: "Being educated is the only way to be free" (Martí 1975, Tomo 8, 289).

2.3 Early Cuban Advances in Medicine

In the course of the 19th century Cuba boasted important scholars in the fields of medicine and natural sciences, who made decisive contributions to the problems of tropical diseases (Pruna Goodgall 2006). Some of them had studied for some years in Europe. In 1803 the physician Tomàs Romay (1764–1849) introduced the anti-smallpox vaccine. The naturalist Felipe Poey (1799–1891) documented Cuban fauna and in 1877 founded the *Sociedad Antropológica* (Anthropological Society); in the last years of his life he accepted evolutionary theories, abandoning his religious faith (Pruna Goodgall 1999). Alvaro Reynoso (1827–1888) studied in Paris, and applied Liebig's concepts to agriculture, proposing a scientific system based on the physics and chemistry of soils for the cultivation of sugarcane.

Carlos J. Finlay's (1833–1915) story deserves special emphasis, since it anticipates in some sense the present American-Cuban controversies in the medical therapeutic field. When the Ten Year War began in 1868, Dr. Finlay (known to Spaniards as a rebel sympathizer) went to live in Trinidad. He returned to Cuba in 1870, and in 1879 he had the opportunity to work with the first American Yellow Fever Commission. He spent years studying mosquitoes and refining his theories, and dedicated over 70 scientific articles for medical conferences and journals to the yellow fever disease, which had caused thousands of deaths in Cuba. By 1881,

---

[6]A complete copy of "Our America" can be found online at http://writing.upenn.edu/library/Marti_Jose_Our-America.html. Last access March 16, 2016.

Finlay had become convinced that the causative agent in yellow fever was a mosquito, probably a member of the species *Aëdes aegypti*. In 1881, however, Finlay was virtually alone in accepting the mosquito–yellow fever connection. His speech of that year to the International Sanitary Conference in Washington, D.C. fell essentially on deaf ears. In 1900, during the first US occupation of Cuba (1898–1902), a US medical commission led by Dr. Walter Reed went to Havana to study the disease.[7] At first the US scientists did not pursue Dr. Finlay's "mosquito" theories, certain that it was "filth" that spread the yellow fever virus. When all their experiments failed, they began to look over Finlay's 19 years long research. A member of the commission, Jesse Lazear, in agreement with Walter Reed, decided to test Finlay's hypothesis by letting himself be stung by a mosquito. He died as a consequence of the experiment. Reed then took advantage of this, but his final report on the aetiology of yellow fever failed to even mention Finlay's theory and research. In it, he took credit for himself for the discovery of the transmission of the disease. Mosquito control programs were introduced throughout Cuba (and in the Panama Canal zone, where work had stopped due to yellow fever outbreaks and many deaths), and the disease was brought under control. In recognition of Reed's contributions to medicine, the Cuban government appointed him the nation's chief health officer and president of the Superior Board of Health in 1902. It took some years before the scientific community finally acknowledged Reed's fraud and Finlay's priority. It was not until the unanimous approval of the motion presented by the Cuban delegation to the 10th International Medical History Congress, held in Madrid, Spain in 1935, that they recognized that Finlay was the first to scientifically prove that the mosquito *Aëdes aegypti* was the transmitter of the disease. In 1954 the International Congress of Medical History formally and officially acknowledged his contribution to the solution of the yellow fever problem, and a symposium in commemoration to him was held in Philadelphia in 1955 (Yellow fever 1955). Before his death in 1915, Finlay was nominated for the Nobel Prize seven times.

2.4   An Aspect of Subalternity: Early Introduction of Advanced Technologies *Versus* a Delay in Basic Sciences

In the meantime, the first scientific institution had been established in Cuba. Proposals for the establishment of an Academy in Cuba had been put forward as early as 1826 by a series of scholars led by Tomás Romay and Nicolás José Gutiérrez, but they remained for long ineffectual. Finally, in light of the scientific developments discussed above, in 1861 Queen Isabella II authorized the founding of the *Real Academia de Ciencias Médicas, Físicas y Naturales de La Habana*

---

[7]On the following events and controversy: Cirillo (2004).

(Royal Academy of Medical, Physical and Natural Sciences of Havana: Pruna Goodgall 1994, 2003; Clark Arxer1999), the first Academy of Sciences in the Americas (analogous Academies were founded in the US 2 years later, in Argentina 13 years later, and in Mexico 23 years later). The considerable lag that occurred between the early introduction of advanced technologies and the delay in the advancement of science and higher education in Cuba in the 19th century is revealing of the nature of Spanish colonial rule in Cuba, and of the increasing penetration of American economic interests (Baracca 2009). Cuba was not particularly rich in natural resources or ore reserves, nor did it develop important transformation industries, apart from that of sugar cane. The island was a source of added value for goods mainly thanks to its strategic geographical position between Latin America, Europe and the United States.

This role was enhanced by the supremacy of the United States on Cuba's trade since the early decades of the 19th century. In this respect one should remark that Cuba suffered indeed not one, but two subalternities at the same time: the direct one, from Spain, was more detrimental, but that from the United States was to imply, as we shall see, more lasting consequences. Already, as soon as in 1826 the volume of Cuba's trade with the United States exceeded that with Spain of almost a factor three (de la Sagra 1831, 200–205). An authority like Fernando Ortìz (1881–1969), a renowned Cuban historian, anthropologist and ethnomusicologist, emphatically asserts:

> … in 1850 the trade of this country with the United States exceeds that with its Spain metropolis, and the United States definitely assume its natural geographic condition of purchaser market of the nearby Cuban production, but also its privilege as economic metropolis. Already in 1881 the Consul General of the United States in the Havana officially writes that Cuba is an economic dependence of the United States although politically it is still ruled by Spain (Ortìz 1963, 64).[8]

Under these conditions, one can understand that Cuba needed neither the contribution of modern scientific knowledge and higher education, nor of particular technological advances in industrial production, as they were instead required, for instance, in Mexico for the development of some industrial fields, like mining industry, minerals and metals. This permits us to understand certain technical innovations in Cuba, such as the introduction of the steam engine in the *ingenios* for cane manufacture, in spite of the abundant supply of slaves, the development of railways, and the fight against tropical diseases. In fact, the island's strategic position lent it great relevance for communication and information technologies, and facilitated the rapid spread of some of the most advanced technologies of the nineteenth century (Blaquier 2009). Interestingly enough, these technologies were not imported to the island from Spain, but from the United States and Britain, and in

---

[8]Cuba's multifaceted relationships with the United States from the early nineteenth century to the island's semi-colonial status in the early twentieth century is the subject of the work by Lorini (2007).

these fields the country anticipated and out-performed its colonial mother-country, while it lagged behind from the purely scientific point of view.

One may therefore suppose that Cuba's flexibility and openness towards technological innovations has in turn contributed to creating a cultural climate and a fecund material basis for subsequent scientific take-off and development. It is difficult, in fact, to believe that the remarkable advances in science after 1959 could have sprung up without fertile soil.

2.5   The Forging of a National Identity, the Ideas of "Cubanity"

In this connection, some remarks ought to be added regarding the versatility, receptiveness and broad-mindedness of Cuban culture, since these features will turn out to be crucial especially in order to interpret recent scientific developments. Cuba is indeed a peculiar melting pot of ethnic and cultural influences from three continents and civilizations. This shaped and strengthened a peculiar kind of national and cultural consciousness and style. As the renowned writer, Abel Prieto—then president of the *Unión de Escritores y Artistas de Cuba* (Union of Cuban Writers and Artists) and former ministry of culture—expresses the process:

> The formation of a properly Cuban culture was an arduous process, long, hard, of zigzags, setbacks and searches, which accompanied in their avatars the creation of national identity; sometimes preceded it; in others, it was dragged by it. The multiplicity and diversity of its ethnic and cultural components, the fierce resistance of the Spanish metropolis to the independence of Cuba, the crucible of the anti-colonial wars, marked in a very peculiar way the birth and first steps of Cuban identity.[9]

As an expert of the Cuban Afro-American tradition has written,

> The Revolution's national ideology of *cubanismo* claims that a homogeneous national culture has been born out of the hybridity.[10]

Fernando Ortìz (1881–1979) coined the term *cubanía*, or *cubanidad* (Cubanity), insisting on the reciprocal influence that various groups had on each other in the creation of a new national identity (Ortìz 1964). Ortìz developed the original concept of 'transculturation' to account for an interpretation of Spanish and African cultural influences in Cuban national identity that acknowledged the ongoing influence of the customs, traditions, and cultures of all those partaking in scenarios of cross-cultural contact and exchange (Font and Quiroz 2005). As he wrote,

---

[9]Abel Prieto, "La Nación y la Emigración" (I Conferencia), La Habana, April 1994 http:// revolucioncubana.cip.cu/wp-content/uploads/2012/12/conf01.pdf. Last access March 15, 2016.

[10]Paula Sanmartín, 2005, *"Custodians of History": (Re)Construction of Black Women as Historical and Literary Subjects in Afro-American and Afro-Cuban Women's Writing*, Dissertation thesis, University of Texas, Austin, p. 39, https://www.lib.utexas.edu/etd/d/2005/ sanmartind11923/sanmartind11923.pdf. Last access March 15, 2016.

> this process does not consist exclusively in acquiring another culture, … rather, the process also necessarily implies the loss or uprooting of an original culture, which could be termed a partial deculturation, as well as the consequent creation of new cultural phenomena which could be described in terms of a neoculturation (Ortìz 1995, 102–103).

The above mentioned specialist remarks:

> Any of the definitions given of *cubanismo*, the most important ideological force in Cuba, have emphasized both political and cultural (especially literary) aspects. Antoni Kapcia describes it as both "a political search for ideology, articulation and identity that preceded and followed 1959; and a literary search for an individual and collective identity".[11] This populist nationalism already defined the intellectual tradition that originated with Martí, and therefore by considering this figure as creator of the Revolution, Castro is appropriating the same discourse. In fact, populism and nationalism are vividly present in Castro's famous speech to artists and intellectuals in the first years of the Revolution, *"Palabras a los intelectuales"* [Words to the Intellectuals] (1961).[12]

Indeed, this sense of a strong national identity emerges in the most representative Cuban writers. As an instance, the Cuban writer and poet José Lezama Lima (1910–1976), considered one of the most influential figures in Latin American literature, writes in a private correspondence:

> Cubanity does not lie in showy tourist attractions, but in an ineffable underground tenderness, a being-not-being, the waving of the breeze, a certain lack of definition, a mixture of the earthly and the stellar. The most solid Cuban tradition may be looking forward to the future. Few peoples of America have been as determined to leap into the future so violently, with a shock of premonition. That is why there is a certain convergence of the generations. We are all marching towards a goal, somewhat distant and uncertain. This vagueness is convenient, it enriches us because it is limitless. Cuban means possibility, fantasy, fever for the future. We need to spread this character throughout the world.[13]

2.6   The Frustration of US Occupation

The War of Liberation from Spain, carefully prepared by Martí, broke out in 1895. The Cuban army obtained substantial gains, but the intervention of the US, which had been feared by Martí, frustrated the ambition for independence. In fact, the two-fold military intervention of the United States in 1898 against Spain in Cuba and in the Philippines brought an end to Spanish colonial rule,[14] and actually

---

[11] A. Kapcia. "Revolution, the Intellectual and a Cuban Identity: The Long Tradition," BLAR 1:2 (1982): 63–78.

[12] Paula Sanmartín, op. Cit., p. 352.

[13] José Lezama Lima, *Cartas a Eloísa y Otra Correspondencia*, Verbum Editorial, Madrid, 2013, pp. 102–103.

[14] A fundamental work on Cuba's relationship with the two "empires", the Spanish and the American, remains: Pérez (1983). The more recent Pérez (2007), is an investigation on Cuban-US cultural relations from 1850 to 1959.

marked the beginning of the US-foreign politics of intervention in the world, exactly as Martí had foreseen. As an irony of destiny, Spain suddenly passed from a hegemonic, although declining, colonial role, to a subaltern one, a change that is too pertinent to the subject of our present study. As a matter of fact, the double humbling debacle and destruction of the Spanish fleets in 1898, in the Pacific and the Caribbean, by the United States arrived as thunderbolts to the Spanish public opinion. In June 1899 the deputy Eduardo Vicente exclaimed at the Spanish *Cortes*:

> I will never be tired to repeat, leaving aside false patriotism, that we follow the example that the United States has given us. This country has defeated us not only because it is stronger, but because it has a level of instruction higher than our one; certainly not because they are braver. No Yankee has clashed with our fleet or army, rather a machine invented by some electrician or engineer. These ones have won the fight. We have been defeated in the laboratory and the offices, not on the sea or the ground (cited in Turín 1959, 375).

During the military occupation of Cuba by the United States (1 January 1899–20 May 1902), and the following nearly six decades (May 1902–January 1959) of restricted independence of the new Cuban Republic, important changes were introduced in the national education system.

Under the military occupation, Enrique José Varona (1849–1933), a Cuban writer, philosopher, and educator, was appointed Secretary of Education and Fine Arts, and introduced a modernization of the Cuban educational system, based on the supremacy of public over private schools and inspired by modern pedagogical ideas. Varona was well aware of Cuba's subalternity to the United States, and that without technical-scientific development (although without radical social changes, impossible under the US occupation) and the start of a process of industrialization, the objective of real independence was an illusion. In a letter (15 October 1900) he wrote to the Cuban doctor and anthropologist Luis Montané[15]:

> You want to know the spirit that guided me when I undertook the reform of our education institutions. […] I acted in the spirit of legal defense of the people of Cuba; a defence within its possibilities and in the field of the possible […] We have to compete in the field of industries and in the field of sciences with the North Americans. And if we want to avoid being completely cancelled from this field we have to educate ourselves as the Americans do… [I] will transfer the fight to the only battlefield where we can fight. We are dealing with a social phenomenon and the consequences of an unavoidable law. The only way to avoid the possible dangers of these consequences is to become part of the conditions producing this phenomenon.

The so-called *Plan Varona* (30 June 1900) put emphasis on active scientific and technological education, in place of the former emphasis on arts and the humanities, though unfortunately he did not increase the teaching of basic sciences, such as mathematics and physics (de Armas et al. 1984; Altshuler and Baracca 2014). Generally speaking, the organization of the University of Havana followed that of American Universities.

---

[15]We thank Dr. José Altshuler for bringing this quote to our attention.

2.7 Social and Cultural Ferments Under US Rule

The following decades were for Cuba a period of crisis, characterized by a web of economic underdevelopment, government corruption and submission to foreign imperial interests, US intromissions and even further military American interventions. In this situation, the original goals set forth for the University by Varona could not be implemented. In particular, the level of the scientific disciplines in Cuba before the Revolution of 1959 depended on social and political conditions that inhibited the technological and scientific evolution of the country. The majority of the Cuban economic and political elites, as well as foreign powers, exploited the island and had no interest in any kind of autonomous development. This situation lasted for the whole period of Spanish colonial and, in different ways, of US-American imperial domination, during which an elite of sugar producers impeded any real advancement of society, especially as regards scientific progress.

However, the problem of cutting loose from the new subalternity to the US empire that had replaced colonial domination, although in different form, as Martí had clearly foreseen, inspired the most lucid minds, despite the resurfacing of strong annexationist political currents.

A revival of progressive and anti-imperialist movements all over the continent was triggered by the student struggles that broke out in 1918 at the University of Cordoba in Argentina and rapidly spread to labour unions and leftist political parties, carrying strongly progressive, anti-private, anti-military, and anti-imperialist goals. This movement not only led to the radical reform and democratization of Argentinian universities, but constituted an epic of emancipation that opened a heroic phase in the development of Latin American universities.

In Cuba these events produced the development of a radical movement, which started in 1923 at the University of Havana, where students proposed a program of reform that aimed at the eradication of the archaic teaching methods then prevailing, and the dismissal of some professors for their evident incompetence. The full reform program was not achieved, but some of the most incompetent professors were replaced by new ones, often proposed by the students themselves. Among these, the physicist Manuel Gran (1893–1962)—a graduate in architecture, civil engineering, and physical and mathematical Sciences from the University of Havana—was put in charge as substitute assistant professor of the two courses of *Física Superior*. In the following years Gran played a very important role, profoundly renovating the discipline by introducing a rigorous approach marked by solid mathematical foundations, problem solving, and practical experiments (Altshuler 2014). The new standards of rigour and method introduced by Gran strongly influenced the teaching of the subject, both at the university and high school levels. Its range was so broad that it was adopted as a useful first introduction to many scientific and technical topics not covered in ordinary courses.

The 1923 reform movement was the start of what has been called the "critical decade" in Cuba (1923–1933), in which the student movement was deeply involved in the struggle against the bloody tyranny of president Machado, who was overthrown in August 1933. In the same year, further measures aimed at modernizing and updating the teaching of physics were introduced both at the university and the high school level, where a number of well-trained teachers was now available. The courses of mathematics and biological sciences were also modernized.

In 1927 another important scientific institution was established in Cuba, the *Instituto Finlay* (Finlay Institute), having as its institutional duty the training of future clerical workers for the sanitary administration; later on, it developed departments for treating tropical diseases with vaccination (Pruna Goodgall 2006, pp. 224–227). In 1937 *Instituto de Medicina Tropical "Pedro Kourí"* was created.[16]

However, on the whole the situation in Cuban universities[17] remained substantially unchanged until 1959, though in the 1950s the regime of Fulgencio Batista did try to promote some sectors of research, as well as some international collaboration, for instance in nuclear physics. When in the mid-1950s the Atoms for Peace campaign was promoted, programs for the construction of nuclear power plants were proposed in almost every country of the western block, including Cuba. However, nobody in the country was actually trained in the field. An exception was Marcelo Alonso, who took graduate courses in physics at the University of Yale, and started a modest laboratory of Atomic and Nuclear Physics at the University of Havana.

2.8 The Weight of Subalternity. Contrasts in Pre-revolutionary Cuba

Even after the modernizing measures of the 1920s and 1930s the general level of scientific development in Cuba had remained modest. Secondary instruction had reached a fairly good standard, for the sectors of the society that had access to it. For instance, in Cuban high school education the teaching of physics was included not only in the curriculum of those who chose the sciences branch in their final (5th) year, but also in the basic curriculum that had to be followed by all students. In the universities, the courses in physics, mathematics and biology, although modernized and made more rigorous, remained basically limited to the 19th century classical theories (Altshuler 2014; Altshuler and Baracca 2014). In physics, for instance, the courses did not cover the modern fields of relativity theory or quantum mechanics. Indeed, not until the late 1950s did Marcelo Alonso introduce the first notions of

---

[16]Pedro Kourí (1900–1964), was a prestigious Cuban physician and researcher.

[17]Besides the University of Havana, there were the *Universidad de Oriente* (Eastern University) in Santiago de Cuba, that had been functioning unofficially as a private institution since 1947, and was made public in 1949 (Méndez-Pérez and Cabal Mirabal 2014), and the Marta Abreu University in Santa Clara, created in 1952.

quantum and nuclear physics. In the biological sciences, the traditional fields of natural history (zoology, botany, geology) had been updated, but the most recent advances, particularly in the field of molecular biology, were not taught. But, above all, genuine research work was neither performed at the academic level nor required for graduation. The job of higher education was almost exclusively the education of the neo-colonial elite and the preparation of secondary school teachers. In any case, the sound level reached in the basic courses produced a foundation of qualified teachers, as well as good textbooks. The rapid take-off of Cuban sciences after the Revolution would not have been possible without this minimum of scientific infrastructure and basis of trained personal.

Besides this renovation of scientific disciplines, the younger generations promoted a lively and original revival in all cultural fields, including music, literature and the visual arts.

In general, in spite of its explosive contradictions and social inequalities and the discrimination against Blacks, in the 1950s the country was actually not underdeveloped: Cuba ranked second in Latin America for average pro-capita income, and among the first five on the basis of other social-economic indicators. The country also boasted one of the best standards of healthcare on the continent, not very far behind those of the United States and Canada. It ranked 11th world-over and third in Latin America for the number of doctors in proportion to the population, although the situation was decidedly worse in rural areas and especially in the Eastern Province. However, Cuba's health sector was unequal: there was only one university hospital and medical school; the private sector predominated, while the public system was rudimentary; two-thirds of the 6300 physicians lived in Havana (Baker 1975; Feinsilver 1993).

For all this period, the Cuban economy continued to be highly dependent on US foreign investments. Difficulties were looming on the horizon, since these investments were gradually being redirected towards oil and industry, with the result that Cuba fell from its place as first investment market for North American capital to the second in 1940 and third in 1956, after Venezuela and Brazil. At the same time, Cuban entrepreneurs preferred to employ the cheap, unskilled labor of impoverished land-workers instead of investing in costly machines. Consequently, hardly any technical innovations were introduced in Cuba in this period, either by importing machines or by developing them inside the country.

2.9 *Granma* Disembarks the Revolutionary Leaders

Meanwhile, the Batista government became a more and more despotic, corrupt regime. The traditional parties became Batista's accomplices, taking part in governments and in the elections of 1954 and 1958. But by the end of the 1940s and the early 1950s the revolutionary movement and its organization were growing. Although Fidel Castro's assault on the *Cuartel Moncada* in Santiago de Cuba of 26th July 1953 was a failure, he managed to transform the trial that followed into a denunciation of the regime (*La historia me absolverá*, History will absolve me).[18] Released thanks to popular pressure, he went into exile in Mexico, where he prepared for the invasion of the Island.

Then on 2nd December 1956, 82 combatants led by Castro landed in the Eastern Province from an overloaded boat named, *Granma*. They were initially decimated, but the revolution that was not only to overthrow Batista's regime, but also to free Cuba from its condition of subalternity to the US empire, had been started—by barely a dozen rebels, whose leaders were not yet thirty years (Fidel Castro was 29, Ernesto 'Che' Guevara 28, Raul Castro 25, and Camilo Cienfuegos 24).

References

Altshuler J (2014) Mathematics and physics in Cuba before 1959: a personal recollection. Baracca Renn Wendt 2014: 107–113.

Altshuler J, Baracca A (2014) The teaching of physics in Cuba from colonial times to 1959. Baracca Renn Wendt 2014:57–106

Baker EL (1975) Cuba study group. The Cuban health care system and its achievement. Cuba's health system: an alternative approach to health delivery. University of Texas Health Science centre at Houston, Houston, TX

Baracca A (2009) Science (Physics) in Cuba: a lag between technological and scientific development? In: Lorini A, Basosi D (eds) Cuba in the world, the world in Cuba: essays on cuban history, politics and culture. Florence University Press, Florence, pp 81–93

Baracca A, Renn J, Wendt H (eds) (2014) The history of physics in Cuba. Springer, Berlin

Blaquier M (2009) Las tecnologías de información y comunicación en Cuba: mitad del siglo XIX e inicios del XX. In: Lorini A, Basosi D (eds) Cuba in the world, the world in Cuba: essays on Cuban history, politics and culture. Florence University Press, Florence

Chomsky A, Carr B, Smorkaloff PM (2003) The Cuba reader: history, culture, politics. Duke University Press, Durham

Cirillo VJ (2004) Bullets and Bacilli: The Spanish-American war and military medicine. Rutgers University Press, New Brunswick

Clark Arxer I (1999) 138 Años de la Academia de Ciencias de Cuba: Visión de la Ciencia y del Proceso Histórico Cubano. Editorial Academia, Havana

de Armas R, Torres-Cuevas E., Ballester AC (1984) Historia de la Universidad de La Habana, 1728–1929, vol 1, pp 237–365. Ed. Ciencias Sociales, Havana

de la Sagra R (1831) *Historia Económica-Política y Estadística de la Isla de Cuba*. La Habana. 1831

Feinsilver JM (1993) Healing the masses. Cuban health politics at home and abroad. University of California Press, Berkely, CA

Figueroa y Miranda M (1975) Religión y Política en la Cuba del Siglo XIX. El Obipo Espada visto a la luz de los archivos romanos 1802–1832. Ediciones Universal, Miami, Florida.

Font MA, Quiroz AW (eds) (2005) Cuban counterpoints. Lexinton Books, The Legacy of Fernando Ortìz

Lorini A (2007) L'Impero della Libertà e l'Isola Strategica. Gli Stati Uniti e Cuba tra Otto e Novecento. Naples, Liguori

Martí J (1953) Obras completas. Havana, Edición del Centenario, editorial Lex, 1953, 2 vols

---

[18]For the full text, see for instance: https://www.marxists.org/history/cuba/archive/castro/1953/10/16.htm. Last access March 15, 2016.

Martí J (1975) Obras Completas. Tomo 8. Editorial de Ciencias Sociales. La Habana. http://biblioteca.clacso.edu.ar/ar/libros/marti/Vol08.pdf. Last access 15 March 2016

Martínez Pérez L (2006) Los Hijos de Saturno. Intelectuales y Revolución en Cuba. Facultad Latinoamericana de Ciencias Sociales, Sede México, D.F.

Méndez-Pérez LM, Cabal Mirabal CA (2014) Physics at the University of Oriente. Baracca Renn Wendt 2014, pp 247–260

Ortìz F (1963) Contrapunteo Cubano del Tabaco y el Azúcar. Editorial del Consejo Nacional de Cultura, La Habana

Ortìz F (1964) Cubanidad y Cubanía, Published in Islas, Santa Clara, vol VI, no 2, enero-junio, 1964:91–96, http://www.fundacionfernandoortiz.org/downloads/ortiz/Cubanidad%20y%20cuban%C3%Ada.pdf. Last access 15 March 2016

Ortìz F (1995) Cuban counterpoint: tobacco and sugar. Duke University Press, Durham

Pérez LA Jr (1983) Cuba between empires. Pittsburgh University Press, Pittsburgh

Pérez LA Jr (2007) On becoming Cuban: identity, nationality, and culture. North Carolina University Press, Chapel Hill

Pruna Goodgall PM (1994) National science in a colonial context: the royal academy of sciences of Havana, 1861–1898. *Isis* 85(3): 412–426

Pruna Goodgall PM (1999) El evolucionismo biologico en Cuba a fines del siglo XIX. In TF Glick, MA Puig-Samper, R Rosaura (eds) El Darwinismo en España e Iberoamérica. Universidad Nacional Autónoma de México, Consejo Superior de Investigaciones Científicas, Madrid

Pruna Goodgall PM (2003) La Real Academia de Ciencias de la Habana, 1861–1898. Consejo Superior de Investigaciones Científicas, Madrid

Pruna Goodgall PM (2006) Historia de la ciencia y la tecnología en Cuba. Editorial Cientifico Técnica, Havana

Quiroz AW (2006) Martí in Cuban schools. In: Font Mauricio A (ed) The Cuban republic and José Martí. Lexington Books, Lanham, pp 71–81

Rodríguez J (1944) Vida del presbítero don Félix Varela. Prólogo de Monseñor Martínez Dalmau, obispo de Cienfuegos. Arellano, Habana. Biblioteca de Estudios Cubanos, 2

Strong AC (2007) The natural education of our America: Jose Marti's philosophy of education. Southern Illinois University at Carbondale. http://gradworks.umi.com/14/44/1444451.html. Last access 15 March 2016

Torres-Cuevas E (1995) Félix Varela: los orígenes de la ciencia y con-ciencia cubanas. Ed. Ciencias Sociales, Havana

Turín Y (1959) L'education et l'ècole en Espagne de 1874 a 1902. Presses Universitaires de France, Liberàlism et tradicion. Paris

Yellow fever (1955) Yellow fever: the complete symposium. Yellow fever, a symposium in commemoration of Carlos Juan Finlay. The Jefferson Medical College of Philadelphia, 22–23 September 1955. Paper 12. http://jdc.jefferson.edu/yellow_fever_symposium/12. Last access 15 March 2016

Chapter 3
Addressing the Challenge of Scientific Development: The First Steep Steps of a Long Path

I do not conceive of any manifestation of culture, of science, of art, as purposes in themselves. I think the purpose of science and culture is man. [In G. Barry Golson ed., *The Playboy Interview*, Interview with Fidel Castro, New York, Playboy Press, 1981, 254]

Abstract Notwithstanding the extremely difficult overall situation and the US blockade and aggression, from the very outset of the victory of the revolution the youthful Cuban leadership showed amazing lucidity and tenacity in their resolutely determination to develop the education, science and health spheres. Their conscious though admittedly ambitious goal was to prepare "a future of men of science" for Cuba. This effort started with a widespread literacy campaign, including the universal right to free education at all levels and a university reform conceived so as to foster scientific research. Seeking and welcoming every source of support and collaboration, from both Soviet and western scientists and institutions, and resorting to their typical inventiveness, from the early 1960s on the Cubans succeeded in laying the foundations for advanced scientific development. In determining the path of this development, every effort of the Cuban leadership and scientific community was driven by the primary purpose of meeting the basic economic and social needs of the country, freeing it from the chains of underdevelopment. The outcomes of these choices were to emerge with surprising swiftness, not only in fields of immediate impact, such as medicine and health, but also with long-term strategic foresight regarding what would be required for future development.

Keywords Scientific development · University reform, 1962 · *Escuela de Física* · Students in the USSR · Collaboration with the USSR · Western physicists at the *Escuela de Física* · Cuban Academy of Sciences · National Centre for Scientific Research · Health care in Cuba

© The Author(s) 2016
A. Baracca and R. Franconi, *Subalternity vs. Hegemony, Cuba's Outstanding Achievements in Science and Biotechnology, 1959–2014*, SpringerBriefs in History of Science and Technology, DOI 10.1007/978-3-319-40609-1_3

3.1 A Future of Men and Women of Science

On January 1, 1959, the revolutionary army entered Havana and the dictator Fulgencio Batista fled to the United States. At that time Cuba had a population of barely 7 million inhabitants, and was an eminently rural country, with scarce natural resources. Foreign interests, in particular North American ones, heavily controlled its economy. Needless to say, the revolutionary leadership faced enormous problems of every kind. In particular, the official literacy rate was between 60 and 76 %, largely because of lack of educational access in rural areas and a lack of instructors (Kellner 1989, p. 61).

In the face of this great challenge, the young revolutionary leadership had a clear consciousness of the link between culture, power and development. In 1961 a massive literacy campaign sent "literacy brigades" to every corner of the country. It was a remarkable success, raising the national literacy rate to 96 % and forcing contact between sectors of society that would not usually interact. As Fidel Castro put it while addressing the literacy teachers,

> You will teach, and you will learn (Serra 2007).

But these actions were part of a more general strategy, aimed not only at the general development of the country, but also at the much more ambitious objective of overcoming the condition of subalternity and reaching real autonomy. To this purpose, the revolutionary government adopted the strategic goal of the development of Cuban science and the construction of an advanced scientific system. Such an ambitious aim might have seemed unrealistic, considering the backward conditions of this small country: but what the young leadership did not lack was courage, since it had undertaken the revolutionary campaign against a regime strongly supported by the United States with barely a dozen guerrillas!

Soon after the victory of the Revolution, in January 1961, President Fidel Castro made his first bold science policy statement,

> The future of our country has to be necessarily a future of men [and women] of science, of men [and women] of thought because that is precisely what we are mostly sowing; what we are sowing are opportunities for intelligence (Castro 1960).

It should be stressed that this utterance was deeply rooted in the tradition of the Cuban freedom seekers, as we have seen in the previous chapter. Such a bold statement was not pure rhetoric, as the subsequent developments of Cuban science were to demonstrate.

Ernesto "Che" Guevara even foresaw at this early time the future importance of solid-state electronic devices and large-scale developments of automation (Pérez Rojas 2014, 282).

> After the 1959 revolution, Cuba made it a priority to find new ways to care for a poor population; part of the solution was training doctors and researchers (Starr 2012).

This has been the cornerstone of Cuba's scientific development ever since.

What is more, this development has taken original paths, largely independent of predetermined models. Despite its lack of tradition and experience, Cuba has been receptive to very different contributions and approaches, and has integrated them with local resources, often with the typical Cuban ability to create an original process of construction of a sound, advanced scientific system.

The lucid and resolute project of the young revolutionary leadership catalysed a collective will in all the components of Cuban society who had chosen to remain with the revolution which boosted forces, and transformed into a *hegemony*—in Gramsci's words (Sect. 1.2), an "intellectual and moral leadership"—over the whole Cuban society.

3.2 Free Education

The goal of offering free education to the population was given highest priority from the very beginning, and it was one of the first steps taken by the new revolutionary government. After the literacy campaign, a campaign of on-going adult education was also undertaken, along with a program to develop an advanced school system open to one and all that had no equal in Latin America. As early as 20 December 1959, the first *Reforma Integral de la Enseñanza* (Integral Reform of Education) was promulgated (Wylie 2010, p. 82). Sixty-nine army barracks were transformed into schools, over three thousand new schools were built in the first year and about seven thousand teachers were trained, with the result that three hundred thousand children could attend school. The doors to secondary and university education were opened to workers, including farm workers.[1]

The universities had been closed since November 1956 because of the brutality of police repression of the students. The generally high standard of the teaching of scientific disciplines at a basic level had made both professors and graduates available for the new university. Their number was however rather limited, and they were further decimated by the emigration of many of them after the victory of the Revolution. These factors delayed the development of a new generation of well-trained scientists. The number of physicists in the country ranked in the order of dozens. The generation that had modernized the field in the past two decades had already disappeared or was no longer active: Manuel Gran was appointed for one year as Ambassador in Paris, and died in 1962; Marcelo Alonso left the country in the end.

At that time, Cuban universities had just over 15,000 students, most of them enrolled in humanities degree courses. In the first decade, 1959–1970, enrolment rose by only 10,000 students, mostly because of the greater opportunities offered to the relatively few students who had completed higher secondary education.

---

[1]See e.g. the interview to the emeritus professor of physics of the University of Havana, Melquíades De Dios: Olimpia Arias De Fuente, An interview with professor Melquíades De Dios Leyva, December 2008, in Baracca et al. 2014b, pp. 285–288.

However, this modest increase was accompanied by a substantial change in the enrolment structure, which now favoured scientific and technological degree courses. The real quantitative leap forward took place in the following decade, when the wave of educational growth that had begun with the 1961 literacy campaign reached the universities, thus increasing enrolment to 155,000 students (MES 1997).

3.3   University Reform, Fostering Scientific Research

From its beginnings, the Revolutionary government developed broad action to foster the growth of Cuban science and technology and the construction of an organic scientific system (Baracca et al. 2006, 2014a, 123–146). In 1961 was created the *Consejo Superior de Universidades* (Higher Universities Council), in which the three then existing universities (in Havana, Santa Clara, and Santiago de Cuba) were represented. The Council laid the foundations for a radical reform of higher education in the country, with free enrolment for all eligible students and a strong emphasis on the development of scientific research. The animated social situation was characterized by a wide gamma of forms of participation, and elaborating this reform involved many social actors. Besides university professors and outside professionals, the student movement, which had actively taken part in the Revolution, played an important role. It took part not only in determining the basic lines of the reform but also in the concrete renovation of the university structure and plans of study and even, as we will see, in teaching activities, to make up for the shortage of teaching staff. After a lengthy debate, the enactment of the Higher Education Reform Law in 1962[2] represented a crucial breakthrough, contributing to the stabilization of the on-going educational process at a time of violent external aggression. This law laid the bases for the development of a modern scientific system, in which teaching was strictly related to scientific research. A few sentences from the Reform Law well illustrate its renovating spirit:

> In today's Cuban society the University is the link through which modern science and technology, in its highest expressions, should be put to the service of the Cuban people; and one of its main goals was to carry out scientific research, develop a positive attitude towards research among university teaching staff and students, and collaborate with scientific institutions and technical organizations beyond the University system.

The Reform established many new degree courses (a 5-year *Licenciatura en Física*, degree course in physics, among them) lacking in the old curricula and needed for the country's economic and cultural development. There was to be a

---

[2]Consejo Superior de Universidades 1962. *La reforma de la enseñanza superior en Cuba*. Havana: Colección Documentos.

Facultad de Ciencias (Faculty of Sciences), which would incorporate seven
Escuelas (Schools): those of Mathematics, Physics, Chemistry, Biological Sciences,
Geology, Geography, and Psychology.[3]

3.4   Early Student-Led Updating of the Teaching of Physics

When the universities reopened their doors in January 1959, courses continued to
be taught essentially in the same way as in the 1950s. However, expectations and
concrete needs pressed for deep change. It was clear that a profound reform of the
curricula and of teaching methods was urgently needed to develop a modern
mentality, one aimed essentially at engaging in scientific research. Since many
former professors had left the country, to cope with the teacher shortage the Faculty
of Sciences hired some high school teachers who not only were graduates in
Ciencias Físico-Matemáticas, but had also acquired on their own a sound knowledge of classical physics. The student movement was active in the process of the
renovation of the university structure and plans of study, and also in teaching.

Although physics was not, and still is not, the quantitatively predominant scientific field in Cuba, its importance emerged in connection with certain pressing
issues faced by the revolutionary government. Among these, Cuba's critical situation urgently demanded the development of a system of short-wave radio communications, and this in turn required the special training of highly qualified
personnel, such as engineers. This triggered an early experience of profound renovation and modernization in the scientific sector that took place in 1960 in the
Engineering School of the University of Havana (which considered communications a vital sector of the new society). The teaching of electrical engineering was
modified to deal with the fields of electronics and electrical communications for
those 4th and 5th year students who agreed to shift to that area, which had been
very poorly covered in the old curriculum. But a radical modification of the
teaching of the basic sciences, mathematics and physics, was also needed, since the
existing four semesters of *Física Superior* did not cover quite a few subjects, now
viewed as necessary to ground the newly introduced engineering subjects. The
students played a direct role in this process. Disregarding copyright laws, a text-
book taken from the most advanced American physics textbooks was prepared, but
serious difficulties were met in teaching their content. The task was entrusted to a
few senior high school teachers and some of the brighter engineering students. After
a truly difficult start, the courses were normalized and incorporated into all the

---

[3]Previously, the Faculty of Sciences was subdivided into three Sections: *Ciencias Físico-
Matemáticas, Ciencias Físico-Químicas,* and *Ciencias Naturales,* whose task was the preparation
of High School teachers.

Engineering majors (Altshuler 2006; Baracca et al. 2014a, p. 133–134). Articles dealing with certain topics of physics of particular interest to engineers were either reprinted or translated into Spanish, and made available to students.

3.5 Students to the Soviet Union

Meanwhile, in the context of the new plans for the industrial development of the country promoted by Ernesto "Che" Guevara, then head of the Industrialization Department of the *Instituto Nacional de la Reforma Agraria* (INRA, National Institute for Agrarian Reform), an important place was allocated to electronics, especially microelectronics.

At the request of the Revolutionary Government, one hundred full scholarships were offered by the USSR for Cubans to study engineering and economics in Soviet higher education institutions. However, only 85 young people could be found who met the academic and other requirements. They left for the Soviet Union in February 1961. While initially no one was supposed to do a degree course in physics, it happened that six months later six of them proposed to major in physics, and the change was approved by Guevara. Later on, a few of their classmates also shifted from engineering to physics. Upon their return home, some of these physics graduates were to make a lasting impact upon the upgrading of physics teaching and research in the country's universities, especially the University of Havana. In the following decades thousands of Cubans were to study, graduate, and get a PhD in the USSR and other countries of the Eastern bloc.

3.6 Fostering Research in Physics as a Strategic Choice, Taking Advantage of All Sources of Local and Foreign Support

The development of a strong and multi-purpose physics sector was considered a strategic priority, a preliminary for all other scientific and technical fields. This choice proved correct, since physics has acted as the backbone of intellectual life and has provided many other fields with scholars, methods and scientific approaches that have been important for the further development of other disciplines in Cuba, such as medicine, biotechnology and the nanosciences. In some ways, the priority assigned to physics was consistent with the post-war international climate: for better or for worse, physics had brought striking advances during the war, and in the rhetoric of the leading countries, including the Socialist ones, it promised miracles, both in itself and as a basic resource for extraordinary progress in every field. But the idea that Cuba's choices conformed to the dominant stereotypes would be misleading, considering the choices made subsequently, which we shall see. A feature peculiar to Cuba was the effort constantly made, from the very beginning, to orient the development of physics towards the priorities needed for the country's social and economic development. The *Escuela de Física* of the University of Havana, which was created de facto in December 1961, faced huge problems in trying to upgrade its curricula and teaching methods (Baracca et al. 2006, 2014a, 135–153). There was an enormous lack of laboratories, equipment, information sources, structures, and material resources, as well as of suitably trained personnel. The School coped with these problems in highly original ways, combining native resourcefulness with the pursuit of all possible kinds of international support and collaboration, both from Eastern and western countries. Students who had started their studies with the old curriculum and were already in the final years did not drop out. Their training was redirected to cover modern conceptions of physics with a view to engaging in scientific research. Moreover, students in the final years of their studies were taken as *alumnos ayudantes* (assistant students) to teach the first-year freshmen (Pérez Rojas 2014, 282–283). "Western" textbooks considered most adequate for the purpose were reproduced and made freely available to teachers and students as so-called "Revolutionary Editions", thus circumventing the economic embargo enforced by the United States government. New laboratories and workshops were created. With the scarce resources available, a great effort was made to direct the physics degree course "toward the modern conceptions of physics, placing research above (ordinary) teaching" (quote from a memorandum of the Director of the School, from Baracca et al. 2014a, 136). The university model followed in this phase was still basically "American", except for the fact that the duration of the basic degree courses (*Licenciaturas*) was prolonged to five years.

In the face of this extremely adverse situation, Cuban physicists immediately exhibited amazing versatility in resorting to every possible source of support and collaboration, as well as an extraordinary ability to integrate them, along with all available local resources, into an original process of construction of their own system. While the first Soviet scientists came to bring help and advice and supported the renovation of teaching, an extremely important factor was the active presence of other foreign collaborators of various origins. In fact, starting in 1961 quite a few professors and experts from many Western countries, in particular several from France[4] and others from the United States, Britain, Italy, Argentina, Israel and Mexico, visited the School of Physics for more or less extended periods of time, even years, providing expert advice, teaching advanced courses in matters not treated before, organizing laboratories and workshops, and promoting the first scientific research activities (Baracca et al. 2014a; Veltfort 2014; Pérez Rojas 2014, 283). These Western scientists contributed to laying the foundations for a

---

[4]In France there was a strong left wing tradition in physics, starting from Frédéric Joliot Curie, continuing with Jean-Pierre Vigier, who was an active supporter of communism throughout his life, and played an important role in promoting the collaborations with Cuba. French physicists promoted coordinate actions in support to the efforts by the Cuban scientists, physicists in particular, to promote the technical scientific development of the country, see also Chap. 4.

renovation of teaching and the development of the early research activities (Veltfort 2014). In the context of this collaboration some of them, working with young Cuban physicists, obtained the first germanium diode as early as 1967.[5] The direction of the University and the School appointed various commissions which had the task of analysing the most important national problems in order to orient the development of physics, also through visits to laboratories and production plants (electronic, metallurgic and chemical).[6]

This was only the beginning of the process. Physics as practiced in Cuba has displayed, as the first of the subsequent disciplines, an exceptional plasticity in its capability to resort to every local resource and to adapt to an often improvised technological infrastructure, in its ability to respond to specific social and economic needs, but also in its openness towards diverse traditions and schools of research.

When the Latin American centre for Physics (CLAF) was created in 1962 as an intergovernmental organization for the promotion of physics in Latin America, Cuba became one of its founding members.

3.7 Leaps Forward in Reaction to Ominous Threats

On the other hand, recourse to diverse international sources of advice and support was common to various activities in Cuba. The Cuban revolution for its uncommon characteristics raised a deep interest all around the world, not only in the young generations, but also in the intellectual circles. In 1960 Jean–Paul Sartre and Simone de Beauvoir visited Havana and met Fidel Castro and Che Guevara, economic counselors arrived from Latin American countries and from East Europe (Michal Kalecki from Poland), and the Marxist economist Charles Bettelheim arrived from France (Hamilton 1992). Not to mention the extraordinary experience of the audacious design of the Escuela Nacional de Arte, conceived of by the Ministry of Culture to host 1500 students from Latin America, Asia and Africa. For this project, Cuban architect Ricardo Porro called his Italian colleagues Roberto Gottardi and Vittorio Garatti to collaborate. Workers, teachers and students were actively involved in the initial phases of planning and construction. Unfortunately, the plan proved to be beyond the island's real financial possibilities and so it remained unfinished, although they have been, and are still used.[7]

---

[5]Fernando Crespo, Elena Vigil, Dina Waisman: "Sobre los primeros resultados en diodos de germanio obtenidos por aleación", Conferencia *Química de Oriente*, Santiago de Cuba, February 1968.

[6]Dr. Daniel Stolik, professor of the Faculty of Physics of the University of Havana, and former Director of the previous School of Physics, personal communication.

[7]See for instance Y. Daley, Cuba's lost Art Schools: An American unearths some truly revolutionary architecture. *Stanford Magazine*. September/October 2000. https://alumni.stanford.edu/get/page/magazine/article/?article_id=39904 (last access March 15, 2016). However, in 1999, by initiative of Fidel Castro, the three architects were invited to come back to Havana to finish the

Few other situations raised such strongly contrasting sentiments around the world as the Cuban revolution, deep enthusiasm vied with fierce opposition. The depth of involvement in this critical situation helps to understand the real issue at stake, the seriousness of the challenge, and the high profile of Cuba's choices. No half-measures would do: there had to be either a surge of pride and bold steps, or the country would fall back into a condition of subalternity. The Eisenhower and Kennedy administrations authorized the CIA to devise ways to remove Castro: attempts were made to poison him, anti-Communist groups inside Cuba were actively supported, and a radio station broadcasted slanted news at the island from Florida. Nothing worked. On April 17, 1961, under a plan funded and implemented by the CIA, 1400 Cuban exiles launched what became a botched invasion at the Bay of Pigs on the south coast of Cuba. Castro's regime came off strengthened.

On the other hand, it is also true that the extreme position of the United States against Cuba was not shared by all capitalist countries. For instance, Canada's policy was always at variance with Washington's.

> Canadian governments have perceived Cuba in another light and have engaged with Havana. Canada has never broken diplomatic or commercial relations with Cuba ... As with other members of the Cold War Western alliance, Canada was wary of American economic warfare. Often, this led to sharp disagreements with the United States, and Diefenbaker's[8] actions during the missile crisis seem to serve as a stark example of the divergence over Cuba (McKercher 2012).

Furthermore, one must consider that Cuba was never completely aligned with the choices of the Soviet Union, but kept margins of autonomous action even in the most delicate moments. We shall verify this concretely as concerns the scientific field.

3.8 Another Strategic Cornerstone: Promoting Medicine and Health Care

In these years the foundations were laid for one of the main achievements of the Cuban revolution, the system of healthcare. By the mid-1960s 3000 physicians (almost one half of the pre-Revolution total) had left the island, primarily for the US. In 1962 the *Instituto de Ciencias Básicas y Preclínicas "Victoria de Girón"* (Institute of Basic and Preclinical Sciences) was inaugurated. Here, 15 university

---

(Footnote 7 continued)

building, although priorities had changed, and due to financial shortages the work proceeds slowly: see Arquitectura de la Revolución Cubana: Escuelas de Arte, http://www.taringa.net/posts/imagenes/1100467/Arquitectura-de-la-Revolucion-Cubana—Escuelas-de-Arte.html (last access March 15, 2016).

[8]John George Diefenbaker was the 13th Prime Minister of Canada, serving from June 21, 1957, to April 22, 1963.

professors of medicine who had not abandoned the country resumed their educational activity, together with other teaching and laboratory personnel (Pruna Goodgall 2006, p. 285). However, the need to expand facilities for biomedical research was soon felt. In 1964, the National Cancer Registry was created, and in 1966 the National Oncology and Radiobiology Institute. A national network of oncological units was also set up, guaranteeing radio and chemotherapy throughout Cuba. Oncology was established as a medical specialty. Mass screening for cervical cancer began in 1967 (Lage 2009).

The development of a first-world level health system was one of the strategic cornerstones of the revolutionary government.

From the outset of the revolution, Fidel has made the health of the individual a metaphor for the health of the body politic. Therefore, he made the achievement of developed country health indicators a national priority. Rather than compare Cuban health indicators with those of other countries at a similar level of development, he began to compare them to those of the United States. This was particularly true for the infant mortality and life expectancy rates. Both are considered to be proxy indicators for socioeconomic development because they include a number of other indicators as inputs. Among the most important are sanitation, nutrition, medical services, education, housing, employment, equitable distribution of resources, and economic growth (Feinsilver 2006).

Malaria, diphtheria and polio were practically eradicated during the first half of the 1960s, and increasingly Cuba evolved towards a developed-country health profile. The substantial reduction of mortality rate, especially infant mortality, achieved in this decade,

...started to evidence the relative increases of congenital defects and genetic diseases. Sickle-cell disease (SCD) and other diseases were emerging as health problems and they motivated the interest and political will necessary for their attention (Lantigua Cruz and González Lucas 2009).

Very soon the health situation in Cuba diverged from the common situation in developing countries. In fact,

The poorer countries of the world continue to struggle with an enormous health burden from diseases that we have long had the capacity to eliminate. Similarly, the health systems of some countries, rich and poor alike, are fragmented and inefficient, leaving many population groups underserved and often without health care access entirely. Cuba represents an important alternative example where modest infrastructure investments combined with a well-developed public health strategy have generated health status measures comparable with those of industrialized countries (Cooper et al. 2006).

Moreover,

As early as 1982, the US government recognized Cuba's success in the health sphere in a report that affirmed that the Cuban health system was superior to those of other developing countries and rivalled that of many developed countries (Feinsilver 2006).

One should add that Cuba became one of the most egalitarian societies in the Third World, and acquired in perspective a rather high ranking in the Human Development Index elaborated by the United Nations Development Program (UNDP 2003).

From the very start, another cornerstone of Cuban policy, as regards international relations, was its support to the non-aligned countries movement, which translated into the active development of "medical diplomacy".

Despite Cuba's own economic difficulties and the exodus of half of its doctors, Cuba began conducting medical diplomacy in 1960 by sending a medical team to Chile to provide disaster relief aid after an earthquake. Three years later, and with the US embargo in place, Cuba began its first long-term medical diplomacy initiative by sending a group of fifty six doctors and other health workers to provide aid in Algeria on a fourteen-month assignment (Feinsilver 2006).

3.9   The Cuban Academy of Sciences

The need to expand and coordinate scientific activities led to the establishment of other higher education and research institutes. In 1962 the *Academia de Ciencias de Cuba* (ACC, Cuban Academy of Sciences) was revitalized[9] and assigned new tasks. It played a decisive role in promoting scientific development in several fundamental branches. Meteorology, geophysics, astrophysics, and electronics were soon established as work groups or departments in the ACC and were consolidated as institutes during the 1970s (Clark Arxer 1999; Baracca et al. 2014a, pp. 142–145). The social relevance of the development of a strong meteorological service must be emphasized, since in recent decades it has allowed Cuba to drastically limit the human casualties that worsening (due to climate change) tropical hurricanes have created in the Caribbean countries (and even in the US), in spite of the unavoidable material and economic damages they cause.

From 1962 to 1963, the Ministry of Industries, then headed by "Che" Guevara, established a group of institutes dedicated to research related to sugar production, sugarcane-derived products, metallurgy, standardization, industrial chemistry and mining. All of these were seeds that started to grow into a broad-based scientific culture, which was later to yield important results. Worth of note is that in 1962 the director of the School of Physics of the University of Havana agreed with the Director of Automation of the Ministry of Industries to offer a course on information theory, preceded by a review of basic calculus, differential equations and complex variables. "Che" attended this course (Pérez Rojas 2014, 282).

In 1964 Fidel Castro inaugurated the new "José Antonio Echeverría" Polytechnic University (CUJAE), which encompassed the Schools of Engineering and Architecture of the University of Havana.

---

[9]As we mentioned in the previous chapter, a *Real Academia de Ciencias, Médicas, Físicas y Naturales de la Habana* (Royal Academy of Medical, Physical and Natural Sciences of Havana) was created in 1861. After the establishment of the Republic in 1902, the adjective "*Real*" was eliminated. In 1962 the new Academy for the first time acquired a national dimension and an effective role.

3.10   The National Centre for Scientific Research

This multiplication of research centres and activities posed the problem of creating a general support structure. An important step towards the consolidation of an efficient scientific infrastructure was the creation in 1965, following a proposal of Fidel Castro, of the *Centro Nacional de Investigaciones Científicas* (CNIC, National Centre for Scientific Research), which was linked to the University of Havana but enjoyed ample autonomy and direct support from the government (Pruna Goodgall 2006, 285–288; Baracca et al. 2014a, 158–159). Its main purpose was to promote and support scientific research in all areas, and to develop post-graduate training. A substantial initial investment was made in equipment (an electronic microscope and a mainframe computer, the first ones in the country, were acquired). Besides collaborating with leading research and higher education institutions in the Soviet Union and the socialist countries, the CNIC also made an agreement with the French National Centre for Scientific Research (CNRS) (Memoria 1976–77), and collaborated with Spain and the United States in the area of neurosciences. Its initial activities focused on biology, chemistry and agriculture. For a long time the CNIC was considered the top centre in the country. It became the generator of other important research centres (*Centro de Neurociencias, Centro Nacional de Salud Animal, Centro de Investigación en Genética y Biotecnología, Centro de Inmunoensayo*, see Chap. 4). Biological sciences would come to form a dominant part of the CNIC, where most members of the highest level scientific staff of today's biotechnological, genetic engineering and pharmaceutical centres have been trained, the collaboration with Western biological and biochemical scientists grew, as we shall see in the next chapter.

References

Altshuler J (2006) Para una historia de las ciencias físicas y técnicas en Cuba. Editorial Científico-Técnica, Havana

Baracca A, Fajer Avila VL, Rodríguez Castellanos C (2006) A look at physics in Cuba. Phys Today 59(9):42–48

Baracca A, Fajer Avila VL, Rodríguez Castellanos C (2014a) A comprehensive study of the development of Physics in Cuba from 1959 (Baracca A, Renn J, Wendt H, eds), pp 115–234

Baracca A, Renn J, Wendt H (eds) (2014b) The history of physics in Cuba. Springer, Berlin

Castro, F. (1960) El Futuro de nuestra Patria tiene que ser necesariamente un Futuro de Hombres de Ciencia. National Agrarian Reform Institute (INRA), Havana

Clark Arxer I (1999) 138 Años de la Academia de Ciencias de Cuba: Visión de la Ciencia y del Proceso Histórico Cubano. Editorial Academia, Havana

Cooper RS, Kennelly JF, Orduñez-García P (2006) Health in Cuba. Int J Epidemiol 35(4):817–824

Feinsilver JM (2006) La Diplomacia Medica Cubana: Cuando La Izquierda Lo Ha Hecho Bien. Foreign Affairs 6(4):81–94 (English transl: Cuban Medical Diplomacy: When the Left Has Got It Right http://www.coha.org/cuban-medical-diplomacy-when-the-left-has-got-it-right/. Last Access 15 Mar 2016

Hamilton N (1992) The Cuban economy. Dilemmas of socialist construction. In Chaffee WA, Prevost G (eds) Cuba. A different America. Boston: Rowman/Littlefield, pp 36–54

Kellner D (1989) Ernesto "Che" Guevara (World Leaders Past & Present). Chelsea House Publishers, Library Binding edition

Lage A (2009) Transforming cancer indicators begs bold new strategies from biotechnology. MEDICC Rev 11(3):8–12

Lantigua Cruz A, González Lucas N (2009) Development of medical genetics in Cuba: thirty nine years of experience in the formation of human resources. Rev Cubana Genet Commun [internet] 3(2):3–23. http://bvs.sld.cu/revistas/rcgc/v3n2_3/rcgc0123010%20eng.htm. Last Access 15 Mar 2016

McKercher A (2012) 'The most serious problem'? Canada–US relations and Cuba, 1962. Cold War History 12(1):69–88

Memoria (1976–77) Memoria Anuario 1976–77. University of Havana

MES (1997) Catálogo. Ministry of Higher Education, Havana

Pérez Rojas H (2014) Interview by A. Baracca: the rise and development of physics in Cuba: an interview with Hugo Pérez Rojas in May 2009 (Baracca A, Renn J, Wendt H, eds), pp 279–284

Pruna Goodgall PM (2006) Historia de la ciencia y la tecnología en Cuba. Editorial Científico Técnica, Havana

Serra A (2007) The "New Man" in Cuba. Culture and Identity in the revolution (Contemporary Cuba). University of Florida, New York

Starr D (2012) The Cuban biotech revolution. http://www.wired.com/wired/archive/12.12/cuba_pr.html. Last Access 15 Mar 2016

UNDP (2003) Human development report 2003. Millennium development goals: a compact among nations to end human poverty. Oxford University Press, Oxford

Veltfort T (2014) The beginning of semiconductor research in Cuba (Baracca A, Renn J, Wendt H, eds), pp 353–356

Wylie L (2010) Perceptions of Cuba. Canadian and American policies in comparative perspective. University of Toronto Press, Toronto

Chapter 4
Reaching a Critical Mass and Laying the Foundations of an Advanced Scientific System

> *...quality of life lies in knowledge, in culture. Values are what constitute true quality of life, the supreme quality of life, even above food, shelter and clothing.* [Ignacio Ramonet and Fidel Castro, *My Life: A Spoken Autobiography*, Simon and Schuster, 2007]

Abstract Despite the initially unfavourable situation of subalternity and threats coming from outside, by freely resorting to a wide range of sources of support, in the surprisingly short time of less than 15 years a critical mass of scientists with a solid basic preparation was reached and the foundations were laid for an advanced, organic scientific system. Around the mid-1960s a lively discussion developed involving scientific, intellectual and student milieus as well as foreign specialists, about what directions and choices to take in order to promote scientific growth that would meet the country's basic social and economic needs and promote human progress. While collaboration with the Soviet Union was strengthened, decisive contributions also came from western scientists, for instance in the summer schools held from 1968 to 1973. Particularly relevant was the contribution of western, mainly Italian, biologists in training a school of Cuban geneticists and specialists in other fields of modern biology in which, because of its ideological hostility, Soviet science had remained far behind.

Keywords Socially-oriented scientific choices · Summer schools in Cuba · Collaboration with the University of Parma, Italy Lysenko case · Contributions of the Italian biologists · Genetics and Physics in Cuba · Paolo Amati

4.1 Rapid Achievements in Science

In spite of precarious starting conditions, revolutionary Cuba's efforts to develop advanced scientific knowledge in higher education and advanced research proved not to be overambitious. The determination of the initial measures achieved rapid

© The Author(s) 2016   39
A. Baracca and R. Franconi, *Subalternity vs. Hegemony, Cuba's Outstanding Achievements in Science and Biotechnology, 1959–2014*, SpringerBriefs in History of Science and Technology, DOI 10.1007/978-3-319-40609-1_4

and remarkable results. In the brief period roughly between the mid-1960s and the mid-1970s, scientific curricula were modernized and a critical mass of well-trained professionals in the basic scientific fields was prepared in Cuban universities and institutions. Advanced scientific activity also began, in some instances reaching a rather good international level. A choice of the sectors to be developed with priority was made, as we shall see, considering the country's most urgent development needs above all. The process of development did not have the same pace and outcomes in the various scientific fields.

The Cuban Revolution, with its unconventional international leaning towards the Third World countries, attracted a plethora of initiatives of support and collaboration, in particular in the field of culture and science, through the promotion of a variety of courses.

The priority assigned to the strengthening of physics—mainly at the University of Havana, but then also at the Eastern University in Santiago de Cuba—succeeded in stabilizing an advanced curriculum, in various steps and through partial results. By 1970 more than 100 students had graduated. They were prepared both to continue doing research and to become basic trained personnel for other Cuban institutions. This rapid process took advantage of collaboration with specialists both from the Soviet Union, which was very advanced in physics, and from western countries, which in many crucial sectors were leaders in the development of innovative technologies, and contributed to certain achievements in Cuba. Besides these developments on the island itself, the number of students that went to graduate or specialize in the Soviet Union and other socialist countries grew considerably in the 1970s. What is more, collaboration with leading Soviet scientific institutions was consolidated over the course of this decade.

The situation was different in the case of the biological sciences. In fact, on the one hand the initial level of Cuban universities in the new frontiers of the discipline, such as molecular biology and genetics, was lower. On the other hand, even Soviet science was far from reaching a competitive international level in the most advanced biological sectors, for many reasons. As a matter of fact, in this field the collaboration of western (in particular Italian) scientists, completely independently of the Soviet Union, was decisive for the formation of the first group of well-trained Cuban professionals by the early 1970s.

In any case, we insist, the progress of Cuban scientific development followed highly original paths, strengthening its *hegemony* over society, at variance with intuitive ideas or usual models. Thanks to the synergy between foreign support and advice, the inventiveness and open-mindedness of Cuban scientists, who were able to integrate a variety of approaches, and the full use of all available local resources, an original process of construction of a sound and advanced scientific system was created.

4.2 Participated and Socially-Oriented Discussion of Scientific Choices

In the case of physics, during the 1960s the main developments still concerned the School of Physics of the University of Havana. The 'first degree' course was begun in 1962 with the enrolment of about 70 students.[1] In the first 5 years around 20 students graduated, some of which were employed in several of the new scientific institutions, or other universities. In spite of the important contribution of the foreign visiting professors to modernize the courses, the teaching staff remained unstable. The return in 1966 of the first 6 Cuban physicists, who had obtained their degrees in the Soviet Union, plus one graduated in the United States, contributed to improving the situation (not without some initial clashes with the attitude of the students actively involved in teaching activities), which was subsequently stabilized with the addition of other teaching staff, and the first students to graduate in physics in Cuba.

An original and very revealing process concerned the choice of the research fields to be developed. Physics is quite a wide field, with many specialized sectors, which were growing in number and specialization in the post-war years, also as a consequence of important technical scientific developments made during the War. Some developing countries were influenced by the ideology of the most advanced frontiers of physics. This ideology was cleverly promoted by the United States to implement a strategy that would enable it to exploit all relevant scientific potential or results that emerged in any country, while steering foreign scientific programs towards basic fields of research that could lead to advanced results but were at the same time far from immediate military applications, over which it kept a strict monopoly (see e.g. Baracca 2012). As a consequence, many developing countries decided (in several cases, with direct US support, especially after the 1953 Atoms for Peace campaign) to buy and build nuclear reactors, or accelerators for charged particles. These were choices unlikely to have significant spin-offs or fall-outs for their domestic economies and development. As we have already observed, there were other new fields boosted by war and post-war research, which had very different implications for a program of autonomous development aimed at overcoming the condition of subalternity. In a masterful study (Krige 2006), John Krige has shown that after World War 2 American strategy in supporting the resurgence and re-launching of scientific activities in the new fields in the European countries resulted in a reinforcement of American hegemony (a process that has been described as a "subtle scientific empire-building", Beyler 2007). Suffice it to consider that in the following decades while Europe reached a leading position in high-energy physics, laboratories and accelerators (explicitly encouraged by the

---

[1] Dr. Daniel Stolik, professor of the Faculty of Physics of the University of Havana, and former Director of the previous School of Physics, personal communication.

US),[2] it definitely lagged behind the US in more practical fields, such as electronics, solid-state and material physics, nuclear technology and or biotechnology, which were far more decisive for overcoming technological subalternity.

In Cuba the choice of which fields of physics to promote followed a completely different path, adopted a different approach, and produced a different outcome (Baracca et al. 2014a, pp. 148–153; Baracca et al. 2014b). In fact, around the mid-1960s a very lively debate took place inside the Cuban community of physicists about which research lines it would be best to develop in order to meet the country's basic development needs, according to the long-term strategy aimed at reaching a condition of autonomy. This debate involved not only the professional personnel and students, but also included political and intellectual circles outside the university.

The debate was still in course at the time of the Cultural Congress of Havana (January 4–11, 1968), a great event with world relevance, attended by 483 prominent foreign personalities from over 70 countries. It is worth spending a few words about it here because of its close connection with Cuban strategy up to that period (Silber 1968; Candiano 2013). The Congress was preceded by previous ones held in Havana, the 1966 Tricontinental and the 1967 OLAS (Latin American Solidarity Organization) conferences. It was part of the project of promoting a new worldwide revolutionary current that was separate from the proposals of Soviet socialism. The 1968 Cultural Congress had a strong revolutionary physiognomy, under the aegis of the defense of Vietnam, Cuba and Third World African and Asian countries. It was specifically focused on the role of the intellectual and culture in revolutionary process of national liberation, referring to Fidel's declaration of 1961 (Sect. 3.1) and to others by the recently murdered "Che" (9 October 1967).

The Congress strongly resonated with the growing movements in capitalistic countries, the student revolt in Berkeley, and incipient protests worldwide. Among the main topics of discussion was the ethical-political responsibility of the revolutionary intellectual to link cultural developments to social progress and the needs of the popular masses, in contrast to the individualism and impoverishment of people in capitalistic countries. Several important international physicists took an active part in the Congress (the Italians R. Fieschi, B. Vitale and D. Amati, the French J.P. Vigier and the Soviet B.P. Konstantinov), and their discussion focused on the choices to be made for the development of physics. Their explicit advice was to avoid the mainstream global choice of elementary particle physics and accelerators, and instead to give preference to solid state physics and related topics. This field was in full development after the war, especially in the United States, and it was providing innovative results and industrial applications that were much more accessible and useful for viable technical and economic development. The French

---

[2]For instance, when the decision to build the big high-energy physics laboratory, CERN, in Geneva, was taken in the 1950s, with European international cooperation and with the explicit approval of the US, the French proposal to also build a nuclear reactor there was discarded. On the other hand, we recall that at least in the 1960s and 1970s, an overwhelming percentage of investments of physics laboratories for equipment went to the United States.

physicists also proposed the organization in Cuba of summer schools, which in fact began the next summer (Sect. 4.4).

As a result of this long debate, it was decided to prioritize the development of the physics of solid state matter (or solid state physics) at the University of Havana. Emphasis was put in particular on semiconductor materials and devices, metals, and magnetism. The choice however was not exclusive, other basic fields, such as nuclear physics or optics, were not eliminated: rather, as we shall see in the next paragraph, specially dedicated centres were created in the meantime, mainly in the Cuban Academy of Sciences, where they were moved and developed, while physicists were left a certain degree of freedom to follow their own preferences (as a matter of fact, a small group working in theoretical high-energy physics was also developed, which has attained appreciable results; Pérez Rojas 2014).

In any case, this choice turned out to be successful and correct, as its subsequent brilliant results and spin-offs have shown. In fact, the development of solid-state physics at the University of Havana was surprisingly quick and it had a direct fall-out for the development of the leading emergent applied fields, such as electronics and microelectronics, which Cuba entered with determination. Again with the support of various countries, in the short span of one decade, the country reached remarkable results.

In the meantime, the preparation of physicists had gone on, to the point where there were now enough physicists to meet both the needs of teaching and the request for well-trained personnel in the other recently created institutions within the Academy of Sciences and the Physics Schools at the Eastern University in Santiago de Cuba and the University of Las Villas. In 1970 there was the highest number of graduates, 67 at the University of Havana (two thirds of whom were employed in other centres), and the first 8 graduates at the Eastern University (Baracca et al. 2014a, 147, 154; Méndez-Pérez and Cabal Mirabal 2014): for the first time all the fundamental teaching activities could be performed by national graduates. In the following four years, more than 190 students graduated in physics at the University of Havana.[3] At the same time, the number of students enrolled in physics jumped to 400, while dozens of Cuban students were studying physics in the Soviet Union, East Germany and other European socialist countries. In the first half of the 1970s, some 50 MSc degrees were awarded in physics (Baracca et al. 2014a, 161).

4.3 A Network of Specialized Technical Scientific Institutions

In the meanwhile, other institutions were created, expanded or transformed, in order to develop other fields of scientific activity that were considered important both for their basic relevance and their social value (Baracca et al. 2014a, 155–157,

---

[3]See the previous footnote.

168–174). Research and development work on microelectronics and optics was promoted at the Polytechnic University of Havana, as well as the School of Physics. Work on geophysics, astronomy, and meteorology began at the Academy of Sciences and was extended to nuclear physics in 1968 and to quantum optics, acoustics and field theory from 1970 on. A "Group for Nuclear Energy", created in 1966 in the ACC, in 1969 grew into the *Instituto de Física Nuclear* (IFN, Institute of Nuclear Physics), with Soviet equipment and support. This Institute was initially devoted mainly to the training of nuclear reactor and radioisotope physicists. Nuclear physics and its applications were also studied at the Eastern University in Santiago de Cuba (besides X-ray and metal physics, optics and spectroscopy), the CNIC and Havana's Oncology Hospital. A centre for Metallurgical Research and an Institute for Metrology were founded at the Ministry of Basic Industry, as well as a Central Laboratory for Telecommunications at the Ministry of Communications. Collaboration with and support from leading scientists and institutions in the Soviet Union and other socialist countries also led to the development of other important facilities and services, such as the Cuban meteorological service (crucial in a tropical country, as we have already remarked), as well as to improvements in communications and in seismic, magnetic, and gravimetric detection.

Thus, a network of more or less specialized technical scientific institutions came into being, within which groups of physicists were created to tackle research and development programs that, when put together, covered quite a wide spectrum of the physical sciences. Moreover, towards the end of the 1960s scientific exchange with Soviet universities and institutes of the Academy of Sciences of the USSR began to be institutionalized, basically aimed not at undergraduate training, but at postgraduate training of young Cuban graduates in physics, and at the strengthening of the country's incipient research groups.

A collaboration with the University of Parma began at the end of the 1960s. The origin of this collaboration deserves special mention. It goes back to Andrea Levialdi, a solid state physicist professor at this university (Waisman 2014). He conceived the ambitious project of developing a long-term collaboration between his university and the University of Havana. For this purpose, though affected by lung cancer, he arrived in Havana on 5 November1968. He died in the Cuban capital city on 8 December, while giving a postgraduate course on solid-state physics. During the month in which he was able to develop these activities, he sent notes to his colleagues in Parma asking them to do their best to send Cuba badly needed materials and equipment (Fieschi 2014: see Levialdi's letters in the Appendix). After his demise, his colleagues in Parma managed to establish a "Levialdi Scholarship," thanks to which the first four Cuban students of physics could graduate in the University of Parma. In the following years other economic supports were found, which allowed a total of around twenty Cuban physicists to receive specialized training in this university and the Special Materials for Electronics and Magnetism Institute of the same (Leccabue 2014).

This collaboration intertwined with Cuban participation in the International Centre of Theoretical Physics (ICTP) in Trieste.

4.4 Summer Schools and Achievements in Microelectronics

New forms of collaboration with "Western" scientists also arose at the turn of the 1960s. Following a proposal that emerged during the Congress of Havana, French physicists (among whom, as we already noted, there was a strong Communist component) organized a "French-Cuban Scientific and University Liaison Committee" with a strong collaboration program (see Cernogora 2014, Document 1, pp. 368–372). Between 1968 and 1973 they organized summer schools in Cuba (Cernogora 2014, Documents 2 and 3, pp. 372–77). Up to 57 courses were given in 1970 by 172 lecturers, with an enrolment of over 1300 participants (Baracca et al. 2014a, p. 153). Italian physicists also took part in these initiatives, giving courses on the teaching of physics. In particular, besides giving advanced courses in these Summer Schools, French physicists supplied much-needed apparatus and materials, and introduced advanced techniques.

It was in these courses that, by the end of the 1960s, exchanges with the French specialists produced a crucial change in the direction of research in solid-state electronic devices (planar technology, substituting silicon for germanium: Document 4 in Cernogora 2014, pp. 378–379). The idea emerged of creating in Cuba a fully integrated cycle, which would go from processing the material to producing the device (Baracca et al. 2014a, 164–168; Vigil Santos 2014). So support from the French specialists (Cuban physicists visited France to obtain technical information) was instrumental in the surprisingly rapid development of advanced microelectronics, in which field since the early 1970s Cuba has reached a level comparable to that of bigger Latin American countries with a longer scientific tradition.

By 1970, devices with a higher level of integration and improved commercial characteristics were being produced, reaching the manufacture of medium integration circuits. In 1975 the first silicon solar cells and in 1977 the first programmable logical array were designed and produced entirely in Cuba. When the 4th Latin American Symposium on Solid State Physics (SLAFES) was held in Havana in 1975, it was clear that great progress had been achieved in Cuba in this field in a surprisingly short time.

In the meantime, cooperation with institutions like the Moscow "Lomonosov" State University, the Leningrad "Ioffe" Physical-technical Institute and other teaching and research centres in the Soviet Union and the East European countries was consolidated. A bilateral cooperation plan (Plan CUSO) was also started with Canada in 1974.

The effort devoted to the development of microelectronics was a wise choice at that moment, since it was an accessible, low-cost technology. In fact it was also pursued in several other developing countries in Latin America and elsewhere. However these programs, successful as they were, were frustrated by unexpected developments. In fact, around the mid-1970s these activities suffered a profound crisis as the result of an unpredictable rapid progress in high integration microelectronics, which only industrially developed countries could afford. Consequently, some activities had to be reoriented, in particular towards opto-electronic sensors and other electronic sectors (Chap. 5).

4.5   Overall Progress in Physics

Remarkable advances were also made in other branches of physics. In 1974, work in nuclear physics was consolidated at the *Instituto de Investigaciones Nucleares* (ININ, Institute for Nuclear Research) and in 1979 the *Secretaría Ejecutiva para Asuntos Nucleares* (SEAN) was created.

Meanwhile, the ACC continued to reinforce and expand its scientific activities. The early work groups on meteorology, geophysics, astrophysics, and electronics were changed into specific institutes. In collaboration with the Soviet Union, they combined basic research with investigations directly related to the Cuban economy in fields like communications, seismic risk, and magnetic and gravimetric maps.

In 1974 a new research unit was created to deal with areas that had not yet been covered in other institutions, the *Instituto de Investigación Técnica Fundamental* (ININTEF, Institute for Basic Technical Research), a multi-disciplinary research unit grouping various departments (ultrasonics, high precision time keeping, remote sensing, solar energy, electric networks and holography). The ININTEF developed important scientific collaborations with corresponding institutes in Czechoslovakia, Poland and the USSR (Baracca et al. 2014a, 172–173). The structure of this Institute was flexible, and it generated other centres, such as the *Instituto de Energía Solar*, established in 1982 in Santiago de Cuba. In 1974, the first Cuban laser was built at the *Instituto Técnico Militar* (ITM).

Within the framework of the Intercosmos Program,[4] in which Cuba had participated since its beginning in 1965, remarkable work was done in communications, meteorology, remote sensing, space medicine and biology, and space research, in close collaboration with the Soviet Union and the East European countries. A high point was reached in September 1980, during the orbital flight of the first Cuban cosmonaut, Arnaldo Tamayo, when roughly 20 scientific experiments, prepared by Cuban specialists from the School of Physics, the ININTEF and the ITM, in collaboration with participants in the Intercosmos Program, were performed in space (Altshuler et al. 2014).

There is a peculiar general feature of Cuban science that should be mentioned, since it is at variance with the dominant worldwide scientific attitude, although it

---

[4]The Soviet Union was the first among space-monitoring countries to follow the path of international cooperation. In 1965 the Soviet Academy of Sciences established a council on international cooperation in the field of exploration and use of outer space. In subsequent years a large-scale, complex program of space research, Interkosmos, was started jointly with socialist countries and some non-aligned nations. It was designed to give nations on friendly terms with the Soviet Union access to manned and unmanned space missions.

has been partially attenuated in the past two decades. The main vehicle of scientific information and scientific results and progress is constituted by articles published in international journals. Moreover, these papers represent a tool of academic recognition and promotion. In recent decades, "Citation Indexes" in each discipline have also been created, and the number of times in which a specific article, or author, are cited in other publications is a title of merit even more important than the number of publications itself. However, as we have already pointed out, in Cuba social goals were given exceptional importance, and so the main value that could be attributed to a scientific contribution was its social benefit and utilization. It is no wonder, therefore, that at least in the first two or three decades, the number of publications by Cuban scientists ranked relatively low (Baracca et al. 2014a, Appendix, Table A.2, p. 229). This attitude was partially attenuated in the course of the 1980s. In recent decades, after the collapse of the Soviet Union, Cuban scientists have increasingly interacted and promoted exchanges (even out of necessity) with the international scientific environment (Chap. 5), accepting therefore the western scientific merit system and being consequently more motivated to publish their results [bibliometric analyses of Cuban publications in physics are presented by: Marx and Cardona (2014), and Altshuler (2014)].

4.6   The Decisive Italian Support to the Development of Modern Biology in Cuba

As we have already mentioned, developments in the biological sciences in Cuba were slower and followed a different path. On the one hand, while the new theories in physics traced back to the beginning if the 20th century and were consequently already well established, the modern innovations of molecular genetics and biology had started in the 1940s, and the field of biological sciences was undergoing deep changes. On the other hand, and even more relevant for the case of Cuba, the Soviet Union, for ideological reasons, had refused modern genetics, and had remained completely excluded from the new developments.[5] For these reasons the Soviets

---

[5]We recall that the origin of the backwardness of Soviet biological sciences has to be traced back to the so-called "Lysenko case". In fact, genetics in the USSR had during the 1930s an internationally appreciated scholar, Nikolai Vavilov (1887–1943). However, around mid-1930s a Soviet agronomist, Trofim Denisovič Lysenko (1898–1976), disputed classical genetics, holding a theory (with a Lamarckist mark), in keeping with Marxist concept of dialectical materialism, along which acquired characters, from environmental and nutrition changes, could become part of the hereditary of a species characteristics and be transmitted to progeny. Lysenko promised rapid progress in Soviet agriculture, violently clashed with Vavilov, and succeeded to expel him from the presidency of the Soviet Academy of Agricultural Sciences. He was strongly supported by Stalin, while Vavilov was prosecuted for plot and sabotage, and died in jail the day before his death sentence was performed. Lysenko's theories were not confirmed and gradually declined, but only after Stalin's death the hegemony of these theories definitively declined, leaving, however, a deep delay in Soviet genetics.

had no influence and role in the developments of Cuban biological sciences, whose absolutely remarkable achievements entirely relied on contributions from western (principally Italian) scientists, and appear all the more relevant.[6]

The early autonomous Cuban efforts in biomedical investigation were originally developed by a group of young physicians at the CNIC, whose initial task was to improve the formation of young graduates. By 1966 the centre had recruited a considerable group of new graduates, many of whom were to be instrumental in the subsequent development of Cuban biotechnology (Pruna Goodgall 2006, 285). Since 1970, work on immunology has gone on in the *Instituto de Oncología y Radiobiología* (Institute of Oncology and Radio-biology) of the Ministry of Public Health.

The decisive leap dates back to the already mentioned Summer Schools promoted by French scientists, starting in 1968: a course of molecular biology was organized at the CNIC by an Italian-Argentinian geneticist residing in Paris, Mario Luzzatti, (which in particular the director of the group of molecular genetics of the International Laboratory of Genetics and Biophysics of Naples, Italy, Paolo Amati, took part in). One of the creators of these theoretical-practical summer courses of biochemistry was Guido Di Prisco of the University of Naples (Italy): after discussions with the Cuban professors of the School of Pharmaceutical Chemistry of the University of Havana, together with Sancia Gaetani, other professors, and a technician of the University of Rome, organized a theoretical-practical course on nutrition, which had a remarkable success and lasted several months.

The summer courses went on until 1973, but the main influence in the formation of Cuban biologists shifted on other, multiple and more systematic initiatives undertaken by Italian biologists.

An early extraordinary contribution was given by an Italian chemist, Bruno Colombo (1936–1989), who in 1969 decided to move from MIT to the Institute of Haematology of the Hospital William Soler in Cuba, which soon became well-known all over Latin America. He then became a molecular biologist, specialized in haemoglobin, and spent 8 years working in Cuba, where his collaboration was fundamental to taking Cuban research in the field to an international level. He also participated in the courses organized by Amati as a teacher of human genetics. Thank to his international connections, he managed to have equipment, reagents and documentary research sent to him (sometimes even from the US), circumventing the embargo. He was so highly regarded in Cuba that he was sometimes sent abroad with official delegations. In 1977 he decided to return to Italy, but he continued to collaborate with Cuba until his death in 1989. He felt

---

[6]The initiatives and the contributions that we discuss in this paragraph have been reconstructed through contacts with various witnesses and some documents that have provided us: the Italian biologists Paolo Amati, Marcello Buiatti, Sancia Gaetani, Luciano and Marina Terrenato, Paola Verani, and moreover Colombo's widow, Maria Cristina Fernandez Lacret, and Gisela Martínez. The latter, was Colombo's collaborator, and successor at the direction of the Department of Biochemistry (subsequently Department of Molecular Biology) of the Institute of Haematology in Havana.

it necessary to write two specialized monographs specifically for Cuban haematologists (Colombo et al. 1981, 1993). His colleague, the chemist Sandro Gandini, also spent a long time working in Cuba.

Coming back to the other initiatives, one must not dismiss an early important contribution to the development of modern genetics in Cuba given by one of the future most important and original American geneticists, James Shapiro. Then a young PhD student involved in the antiwar movement, he left Harvard and went to Havana, where he taught at the School of Biological Sciences from August, 1970 to April, 1972.[7]

But the more important and long lasting initiatives were promoted by Paolo Amati. When he took part in the 1968 Summer School, he realized that the lack of knowledge of Cuban biologists in the modern field of genetics, which he considered crucial for the development of molecular biology in Cuba, would require more in-depth treatment. Therefore, in 1969 he proposed to the rector and the vice-rector of the University of Havana, a six-month course specifically focused on genetics and statistics, picking some twenty Cuban students from universities and research centres, with the aim of creating an environment of well-trained geneticists in Cuba. Amati also asked to visit the centres potentially interested in the proposal, in order to adjust the course to the specific needs of Cuba. He was accompanied by Luis Herrera, a student of the first 1968 summer course whom Amati met again in Paris where he was studying genetics, and who was to become the director (up to 2015) of the Centre of Genetic Engineer and Biotechnology (Chap. 5). The first postgraduate six-month course organized by Amati took place from January to June 1971. It was divided in two parts, three months dedicated to genetics and biometry, and three months to human, microbe, animal and plant genetics respectively. The students of the course were biologists, veterinarians, agronomists, and physicians, chosen among the most promising candidates. The teachers included some of the best Italian specialists. The participants were divided into four groups. In this phase laboratory activities were planned. At the end of the course, the Cuban students were expected to begin, under the direction of the professors, research work on topics proposed by the University of Havana or other Cuban institutions. Amati also gave the Cubans an Olivetti 101 computer, which was especially useful for genetics.

The course was then expanded, after discussing the results with the participants, and it was repeated for two more years, gradually involving the Cuban students in its organization and as teachers. In subsequent years exchanges continued, with several Cuban biologists visiting the Italian colleagues and laboratories. These courses were extremely important, since the students later went on to make up the backbone of the Cuban biomedical sector.

---

[7]James Shapiro, Curriculum Vitae, http://shapiro.bsd.uchicago.edu/cv.shtml. Last access March 15, 2016.

... a similar pole [of human genetics, arose] at the Faculty of Medical Sciences in Santa Clara province, supervised and directed by a biologist, Dr. Klaus Atlas, biochemistry professor at the University of Giessen in the Federal Republic of Germany ... The teaching of Genetics in undergraduate courses began immediately after ending Professor Amati's graduate course and, from the 1971-1972 academic years, it took place simultaneously with the formation of the first specialists (Lantigua Cruz and González Lucas 2009).

But these were not the only initiatives. In fact, in the summer of 1970 the Italian virologist Giovanni Battista Rossi (1935–1994) taught a course in Cuba. The following year Paola Verani repeated his virology course (and was also involved in the study of the mosquito transmitting dengue illness), and Rossi taught a third course in 1972. They were both also involved in the 1971 swine flu epidemics (see below). Another course was given by the Italian oncologist Romano Zito. Other summer courses were organized by Italian biologists, specifically devoted to the formations of secondary school professors of biological sciences. Worthy of note is also the first common Italian-Cuban publication, which came out in 1973, on the study of protein composition of milk of a species of Cuban cows (Cervone et al. 1973).

In the meanwhile, thanks to heavy investments in equipment, the CNIC was growing as a centre of excellence for Cuban science, and its neurophysiology and microorganisms laboratories achieved special importance for biology. The groups that emerged from there went on to play a central role in subsequent developments in biotechnology (Pruna Goodgall 2006, 285–288).

Paolo Amati's collaboration did not end with the above-mentioned initiatives. He continued to provide support and advice for numerous scientific developments in Cuba (indeed, he is better known by the name Pablo, thanks to his uninterrupted collaboration).[8] For instance, he helped create a centre of genetics in the Hospital Hermanos Ameijeiras in Havana, taking advantage of international support to equip the centre, and he promoted other international courses. After his transfer from Naples to the "La Sapienza" University of Rome, the Cuban students who had followed his courses and activities repeatedly visited his department there for longer or shorter periods of time.

In 1979 the *Instituto de Medicina Tropical y Parasitología Pedro Kourí* (Institute of Tropical Medicine and Parasitology), which was to become Cuba's leading public health institute (see below), was established in Havana.[9]

The Animal Health Division played an outstanding role during the 1971 and 1980 swine flu epidemic, and later generated a specialized centre for agricultural and zoo-technical health. Until 1977 the main training centres were the Medical Sciences Higher Institutes in Havana (ISCM-H) and Villa Clara (ISCM-VC) (Lantigua Cruz and González Lucas 2009).

---

[8]For his thirty-year collaboration with Cuba, in 1998 the Council of State awarded Pablo Amati the Finlay medal, the highest Cuban honour for scientific merits.

[9]From a previous institute founded in 1937 by Pedro Kourí (1900–1964), a prestigious Cuban physician and researcher; in the 1960s, after Kourí's death, the institute had declined.

4.7 Growing Institutional Planning of the Cuban Scientific System

In brief, the creation of efficient scientific structures, the acquisition of essential equipment and the preparation of a critical mass of trained scientists in the basic scientific fields by the turn of the 1960s, made the early successes possible, as for instance in the case of microelectronics. Moreover, the formation of a group of specialists in advanced biological fields laid the basis for the subsequent exceptional leap (Chap. 5).

At this point, a modern, well-integrated scientific network had been laid, and the Cuban system of scientific research took shape (Baracca et al. 2014a, 175–180). On the other hand, the growing complexity of the system and the multiplication of activities could not be left to the initial almost autonomous initiative, and inevitably required new means of central coordination. This coordination proceeded in successive steps. Initially, it caused some friction among single institutions and groups that had previously enjoyed greater autonomy, however in the long term these groups learned to complement and appreciate each other.

In view of the growing dimensions and complexity of the system, in 1974 the *Consejo Nacional de Ciencia y Técnica* (National Council for Science and Technology) was created. In 1977 its functions were transferred to the ACC, where the *Dirección de Ciencias Básicas* (Direction of Basic Sciences) arose.

As regards higher education, in 1976 the *Ministerio de Educación Superior* (MES, Ministry of Higher Education) was created. The following decade saw a period of successive changes, followed with mixed results before stability was achieved. Higher education curricula and research activities were reorganized, which initially created some problems. Specialized teacher training institutes of higher education were created, called *Instituto Superior Pedagógico* (Higher Pedagogical Institute), with institutes in various provinces.

During the 1960s and 1970s, Cuba trained about 1.8 researchers per 1000 inhabitants, a figure far above the average in Latin America (0.4) and close to that of Europe (2.0) (UNDP 2001).

This situation placed Cuba well outside the trend of correlation between the size of a countries' scientific system compared with that of its economy, and subsequently these favourable conditions permitted a new development program to be established (López Mola et al. 2006).

The spectrum of international scientific collaborations remained wide, well-articulated and open. Collaboration with the most advanced scientific institutions in the Soviet Union and other socialist countries was strengthened and established formally, especially in physics, and in technology and related fields. Thousands of Cubans have studied and specialized in these countries. However, from the 1960s on numerous Cuban scholars have gone to Italy to specialize or take a Ph.D., including approximately 20 physicists in Parma, and biologists, mainly in Rome, as well as to France and other countries. In biology, as we have seen, the decisive collaboration, mainly with Italian scientists, developed starting from the late 1960s. In the following decade this was widened to include North American, Finnish and French specialists and institutions, as we shall see in more detail in the next chapter. It is remarkable that in 1976, at the peak of Cuban-Soviet cooperation, Cuban scientific institutions were also affiliated with forty-eight international scientific organizations, and another thirty-three applications were being processed (MINSAP 1976: cit. in Reid-Henry 2010, 175).

References

Altshuler E (2014) Contemporary Cuban physics through scientific publications: an insider's view (Baracca A, Renn J, Wendt H, eds), pp 439–446

Altshuler J, Calzadilla Amaya O, Falcon F, Fuentes JE, Lodos J, Vigil Santos E (2014) Cuban techno-physical experiments in space (Baracca A, Renn J, Wendt H, eds), pp 293–300

Baracca A (2012) The global diffusion of nuclear technology. In: Renn J (ed) The globalization of knowledge in history, 2012, Studies I: Max Planck research library for the history and development of knowledge. Edition Open Office, Berlin, pp 669–771

Baracca A, Fajer Avila VL, Rodríguez Castellanos C (2014a) A comprehensive study of the development of Physics in Cuba from 1959 (Baracca A, Renn J, Wendt H, eds), pp 115–234

Baracca A, Renn J, Wendt H (eds) (2014b) The history of physics in Cuba. Springer, Berlin

Beyler RH (2007) Review of Krige, John, American Hegemony and the Postwar Reconstruction of Science in Europe. H-German, H-Net Reviews. http://www.h-net.org/reviews/showrev.php?id=13388. Last Access 15 Mar 2016

Candiano L (2013) El Congreso Cultural de La Habana de 1968. Marcha (5 April). http://www.marcha.org.ar/el-congreso-cultural-de-la-habana-de-1968/. Last Access 15 Mar 2016

Cernogora J (2014) A witness to French-Cuban cooperation in physics in the 1970s (Baracca A, Renn J, Wendt H, eds), pp 365–379

Cervone F, Diaz Brito J, Di Prisco G, Garofano F, Gutierrez-Noroña L, Traniello S, Zito R (1973) Simple procedure for the separation and identification of bovine milk whey proteins. Biochim Biophys Acta 295:555–563

Colombo B, Svarch E, Martinez G (1981) Introduccion al estudio de las Hemoglobinopatias. Editorial Cientifico- Tecnica, Ministerio de Cultura, La Habana

Colombo B, Guerchicoff Svarch E, Martínez Antuña G (1993) Genética y Clínica de las Hemoglobinas Humanas. Editorial Pueblo y Educación, La Habana. ISBN 959-13-0044-1

Fieschi R (2014) The Andrea Levialdi Fellowship (Baracca A, Renn J, Wendt H, eds), pp 361–364

Krige J (2006) American hegemony and the postwar reconstruction of science in Europe. MIT Press, Cambridge MA

Lantigua Cruz A, González Lucas N (2009) Development of medical genetics in Cuba: thirty nine years of experience in the formation of human resources. Rev Cubana Genet Comunit [internet] 3(2):3–23. http://bvs.sld.cu/revistas/rcgc/v3n2_3/rcgc0123010%20eng.htm. Last Access 15 Mar 2016

Leccabue F (2014) My collaboration with Cuban physicists (Baracca A, Renn J, Wendt H, eds), pp 381–386

López Mola E, Silva R, Acevedo B, Buxadó JA, Aguilera A, Herrera L (2006) Biotechnology in Cuba: 20 years of scientific, social and economic progress. J Commercial Biotechnol 13:1–11

Marx W, Cardona M (2014) Physics in Cuba from the perspective of bibliometrics (Baracca A, Renn J, Wendt H, eds), pp 423–438

Méndez-Pérez LM, Cabal Mirabal CA (2014) Physics at the University of Oriente (Baracca A, Renn J, Wendt H, eds), pp 247–260

MINSAP (1976) Informe Annual. Havana

Pérez Rojas H (2014) Interview by A. Baracca: the rise and development of physics in Cuba: an interview with Hugo Pérez Rojas in May 2009 (Baracca A, Renn J, Wendt H, eds), pp 279–284

Pruna Goodgall PM (2006) Historia de la ciencia y la tecnología en Cuba. Editorial Científico Técnica, Havana

Reid-Henry S (2010) The Cuban cure: reason and resistance in global science. University of Chicago Press, Chicago

Silber I (1968) Cultural congress of Havana, New York. ASIN: B0007DRHVU

UNDP (2001) Human development report 2001: Making new technologies work for human development. United Nations Development Programme. Oxford University Press, Oxford, UK

Vigil Santos E (2014) Experimental semiconductor physics: the will to contribute to the country's economic development (Baracca A, Renn J, Wendt H, eds), pp 289–294

Waisman D (2014) Andrea Levialdi in memoriam (Baracca A, Renn J, Wendt H, eds), pp 357–360

Chapter 5
The Decisive Leap in the 1980s: The Attainment of Cuba's Scientific Autonomy

> *Cuba's achievements in social development are impressive given the size of its gross domestic product per capita. As the human development index of the United Nations makes clear year after year, Cuba should be the envy of many other nations, ostensibly far richer. [Cuba] demonstrates how much nations can do with the resources they have if they focus on the right priorities—health, education, and literacy.*
> [Kofi Annan, Secretary General of the United Nations, April 11, 2000]

Abstract Further developments in the 1980s led the Cuban scientific system to its maturity, allowing it to achieve striking results in various fields and to demonstrate its substantial autonomy. The physics sector was profoundly reorganized. Nuclear physics and technology acquired pre-eminence thanks to the decision to build a nuclear power plant provided by the Soviet Union. Other sectors were also reorganized, and surprising results were reached in the completely new field of superconductivity. But the most enduring results were achieved with the development of a self-reliant field of biotechnology, just as it was emerging worldwide, and without any aid whatsoever from the Soviet Union. As always, this initiative, strongly supported by Fidel Castro, was prompted by the emergent demands of the health system after the typical third world diseases had been practically eradicated. Interferon technology was learned and quickly reproduced and mastered in the early 1980s through contacts with American and Finnish specialists. Soon after, recombinant DNA technologies were independently developed. In the late 1980s a large industrial scientific complex was built which soon started producing and commercializing Cuban-made medicines and vaccines.

Keywords Superconductivity at high temperatures · Project of a nuclear power plant · *Proceso de rectificación* · Interferon and vaccines in Cuba · Genetic engineering in Cuba · Western Havana Bio-Cluster

© The Author(s) 2016
A. Baracca and R. Franconi, *Subalternity vs. Hegemony, Cuba's Outstanding Achievements in Science and Biotechnology, 1959–2014*, SpringerBriefs in History of Science and Technology, DOI 10.1007/978-3-319-40609-1_5

5.1   New Planning of Scientific Development, with the Goal of Reaching Autonomy

The remarkable advances and concrete results attained in the first two decades after the victory of the Revolution constituted a sound basis for further decisive progress towards the achievement of Cuban science's substantial autonomy. The Cuban scientific system grew further in size, complexity, and fields of activity. New research centres were established and existing branches received additional support. International collaboration agreements and participation in joint programs increased, and involved practically all sectors and institutions in the country up to around 1990 (a turning point, when the collapse of the Soviet Union radically changed the situation).

There was, however, a significant change compared to the previous period: since Cuba had finally adopted a Soviet-style planning system in the economy, decisions about the organization of research activity and higher education, as well as the establishment of new collaboration agreements, were no longer left to the initiative of individual institutions or groups. Rather, as mentioned, the most important decisions were taken over by the ACC. Thus, in a certain sense, an early "romantic" period came to an end. This period had been momentous and, combined with Cuba's characteristic initiative, probably invaluable in boosting Cuban science and higher education in a first phase, under backward and extremely difficult conditions. But although a mature system cannot do without coordination and planning, the originality of Cuba did not cease to play a significant role, even in large-scale projects, as we shall see below regarding some important cases. Moreover, this new dimension did not mean that the basic problems of the country, particularly but not only in the field of health, were overlooked. Applied research was prioritized, much of it related to the learning, adaptation and integration of new technologies to local conditions, though room was allowed for basic research aimed at providing a solid base for the training of specialists and the future development of Cuban technology. In fact, the technologies transferred to Cuba were not always among the most *internationally* advanced, since they entailed relatively high energy consumption, low levels of automation and excessive aggressiveness towards the environment. Still, for Cuba they seemed to represent a considerable leap forward and a great challenge for its young scientific and technological community. And their usefulness for, and adaptability to, the most urgent problems of the country was always a primary criterion of choice.

In the new, planned system, research was no longer organized according to scientific branches, but to spheres of application. Research centres became mainly multidisciplinary, and this has remained a peculiar feature of Cuban science.

In this context, the original choice of favouring the development of a strong physics sector proved correct, since physicists were employed in a wide variety of institutions, where they were responsible for activities not necessarily recognizable as related to physics and where they contributed to the development of rigorous methods and approaches. New research centres, generally formed from pre-existing groups, emerged to strengthen the original collectives by providing them with greater means to apply the scientific results achieved. The Cuban scientific system grew notably in dimension and complexity, as well as in its coordination.

Such developments were all the more remarkable as they marked a sharp countertrend with respect to the rest of the continent. In fact, in the 1980s and early 1990s politics of economic austerity and financial constraint dominated Latin America. Thus, the fact that Cuba continued to increase its investments in science, and in particular in health and medicine, could hardly be more amazing. It is true that Cuba belonged to the socialist bloc, but its policies appear exceptional even in this perspective, the more so since Cuba was to confirm these choices as strategic even in the worst period following the downfall of the Soviet Union (see Chap. 6).

We will now discuss Cuba's developments in the fields of physics, in particular electronics and nuclear technology, and biotechnology.

5.2 Fostering Electronics, and "Improvising" Superconductivity

Investment and development planning had a strong impact on the advancement of the country's electronics and nuclear projects, and to a lesser degree on the branches of physics related to the mining-metallurgical and iron-mechanical sectors.

The crisis of the microelectronics sector following its initial brilliant developments (Chap. 4) called for its re-organization and recovery (Baracca et al. 2014a, 183–189; Baracca et al. 2014b). A National Program for the Development of Electronics was created, under the vice presidency of the Council of Ministers. In 1980 an "Electronic Front" grouped together all the existing organizations in the field: managing institutions for science and technology, industry, institutions of higher education, research centres, technological schools, import firms, software development centres, etc. The independent efforts that Cuba had previously made in the fields of electronics, microelectronics and computing were coordinated, with the purpose of transferring their developments to the country's industry or to corresponding COMECON programs, in which Cuba played an increasingly autonomous role. In a sense, one could interpret Castro's 1961 prophecy (Chap. 2) as having come true, in the sense that science was becoming the most precious resource of this poor small country, and Cuba a reservoir of

men and women of science,

and know-how for both underdeveloped countries and the socialist block.

The design of a plant for manufacturing silicon microelectronics components in the province of Pinar del Río was restructured technologically, and the plant was officially inaugurated in 1987 (equipment was bought from a Spanish firm later absorbed by the US-controlled firm Motorola).

During the 1980s, scores of physicists engaged in the development of electronic devices and equipment, telecommunications, software and robotics at the Central

Laboratory of Telecommunications (LACETEL), at the ACC, the CNIC and other institutes. In order to broaden the scope of academic work in the universities, and to link it more directly to the economic development of the country, higher education institutions had to be supplied with the necessary infrastructures, logistics and management capability. To this end, the *Instituto de Materiales y Reagentes para la Electrónica* (IMRE Institute for Materials and Reagents for Electronics) was created in 1985 at the University of Havana, in order to concentrate resources, attract state investment and give basic support to the National Program for the Development of Electronics (Baracca et al. 2014a, 185–187). It incorporated in a more well-coordinated and effective organization the solid-state physical research laboratories that existed in the physics faculty, and other groups from the chemistry faculty. From 1986 to 1991, the IMRE grew considerably and received investments that allowed it to increase its staff and infrastructure and purchase new research equipment. Two visits by President Fidel Castro, in 1987 and in 1989 respectively, brought additional support to the Institute. The IMRE maintained very close working relations not only with the units belonging to the Electronics Front, but also with several institutes in the public health system and other research and production centres, providing a wide range of services of substantial economic and social value, which proved to be of strategic importance for the country.

The collaboration with the Ioffe Institute in the Soviet Union, especially with the Laboratory for Contact Phenomena led by future Nobel Laureate, Zhores Ivanovich Alferov, became wide-ranging and effective.

A very telling story is the sudden birth, starting practically "from nothing" and relying only on local resources, of some brilliant research activity in superconductivity in 1987 (Arés Muzio and Altshuler 2014). Until then, no experimental research in this field had been carried out in Cuba, since liquid helium was essential for these activities and it was not available at any local institution.[1] At the end of February 1987, Paul Chu announced in Houston, Texas, that he had obtained an yttrium barium copper oxide (YBaCuO) superconducting ceramic whose critical temperature was above 90 K (-183 °C). Chu's discovery was easier to replicate, as it involved cooling the sample with liquid nitrogen, a more accessible technology. Laboratories worldwide were now able to reproduce this finding based on information available about the new materials. As a result of the experience accumulated at the magnetism laboratory with ceramic materials, Oscar Arés Muzio was asked to

---

[1]The phenomenon of superconductivity, discovered by Kamerlingh Onnes in 1911 in Leiden, consists in the electric resistance of certain materials vanishing to zero, and the magnetic field being expelled, below a characteristic critical temperature of a few degrees above absolute zero (-273 °C), precisely below the liquefaction temperature of helium (4 K, that is -269 °C). Work on superconductivity had therefore been reserved to laboratories that had liquid helium, a very expensive refrigerant, at their disposal. In 1986 G. Bednorz and A. Müller broke the temperature barrier, discovering a compound with higher critical temperature (35 K, -248 °C, which was, however, still below the liquefaction temperature of nitrogen, 77.35 K, or -195.80 °C). For this discovery, they were awarded the Nobel Prize. Less than 6 months later, Chu and his collaborators reached a critical temperature above 90 K, a temperature higher than the liquefaction temperature of nitrogen. Further progress has reached critical temperatures of 150 K.

try to create a high temperature superconductor. Under his leadership, barely two months after Chu's announcement a team managed to obtain and produce the first YBaCuO superconducting ceramics in Cuba. This was given widespread coverage in the national media in early May, and generated numerous talks and seminars referring to the new scientific development. It was now clear that the IMRE was capable of engaging not only in solving specific practical problems, but also in obtaining world-standard scientific results. A new group was created which attracted many brilliant young researchers, and was endowed with resources for setting up a new superconductivity laboratory, which received Fidel Castro's direct interest.

This experience, however limited, reveals quite a lot about Cuban science. Starting from no previous experience in the specific field, Cuban scientists entered a highly innovative and advanced scientific field only a few months after its discovery, relying only on local resources, and developing scientific activity at a good international level. But, as we will have to say about other accomplishments as well, this is only the superficial aspect of the story: the really significant aspects are the different reasons, attitude and means guiding Cuban scientists upon entering this completely new field, as compared to leading laboratories around the world. The Cubans were certainly well-trained at the highest international level, but they also possessed special flexibility, open-mindedness, resourcefulness and motivations.

Parallel developments of this apparently isolated effort are also revealing. About the same time, the liquid helium technology was also purchased, since it was necessary for cooling the nuclear magnetic resonance tomograph independently acquired by the "Hermanos Ameijeiras" Hospital, which had been directly authorized by Fidel Castro after a visit at IMRE. The helium liquefying plant was installed at IMRE, which would be responsible for the supply of the required liquid helium, including supplies for IMRE's own research. So the new laboratory was supplied with equipment valued at more than half a million US dollars, which enabled it to take on some ambitious projects.

A few months later, YBaCuO superconducting ceramics were also obtained at the recently created centre for Studies Applied to Nuclear Development (CEADEN), where research on superconductivity had also been started.

5.3 The Project of a Nuclear Power Plant: Nuclear Physics as the Backbone of Cuban Scientific System

Actually, the CEADEN had been created to carry out applied research and development, technology assimilation, technical and scientific services and other activities supporting the national nuclear program, but it also had broader goals. It was officially inaugurated on 28 October 1987 at new premises in the presence of President Fidel Castro and the Director General of the International Atomic Energy Agency (IAEA), Hans Blix, and it housed well-equipped laboratories for solid-state physics, nuclear physics, radiochemistry and nuclear electronics.

The programs in the nuclear field were particularly ambitious (Baracca et al. 2014a, 189–193), especially since in 1976 Cuba had agreed with the Soviet Union to build a nuclear power station in Juraguá (Cienfuegos Province), with the goal of reducing dependence on imported oil. This entailed the introduction of the very complex system of nuclear technologies in the country's economy and the creation of a system of nuclear radiological protection and safety. This, in turn, required a re-organization of the whole sector, which was given a special, favored place and structure, based on a centralized model similar to that of the Soviet Union.

The nuclear program was developed in close collaboration with the USSR, the European socialist countries and the International Atomic Energy Agency (IAEA). Cuban students and researchers gained access to top universities and research centres and to IAEA projects. The program required investment operations, training of cadres, research and collaborations, as well as other initiatives, and so required important political and organizational decisions. Moreover, a nuclear program always requires centralized decision-making, organization and forms of political control. Thus, a new central body was created, the Cuban Commission for Atomic Energy (CEAC), responsible for coordinating and controlling the efforts of the main national institutions involved in nuclear activities and for advising the government on policies in this sphere. Several institutions were created to give scientific and technological support to the program, and significant changes were also introduced at the university level. The profound changes associated with these developments, along with a certain lack of discussion within the scientific community, led to some tension and discontent. On the basis of a technical evaluation of the personnel belonging to the Institute for Nuclear Physics, a group of its employees who were not considered suitable for the job were relocated. This decision was obviously strongly contested, since many thought that it would have been more advantageous to take advantage of the experience accumulated in the pre-existing centres, devoting the resources allocated to the nuclear program to them. As a matter of fact, some of the newly created training and research centres effectively represented an undesirable duplication of efforts and programs.[2]

During the 1980s, hundreds of the best Cuban high-school graduates were awarded scholarships to specialize in nuclear sciences and technologies, including nuclear physics, in the USSR and other socialist countries. To guarantee adequate preparatory training for students, fourteen Exact Sciences Vocational Senior High Schools (IPVCE) were created in the country (one per province), which to this day have been the main source of Cuban students for science degree-courses in national universities.

---

[2]One of the most visible examples of this was the creation in 1981 of a Faculty of Nuclear Sciences and Technologies (FCTN) next to the main campus of the University of Havana, and later of the Higher Institute of Nuclear Sciences and Technologies (ISCTN), for the purpose of teaching degree courses in nuclear physics and radiochemistry. This meant that an elite group of physicists, chemists and engineers with a different and more specific profile was being trained.

5.4   Redirecting Scientific Development

These developments were also a result of the *Proceso de Rectificación* (Process of Correction) promoted by the Cuban government around the mid-1980s, which was not only a re-organization needed due to the growing complexity of the system, but also a reaction to both the *glasnost* promoted by Gorbachev and Reagan's more aggressive foreign policy. This process involved a re-centralization of the economy and a return to a strong ideological emphasis on the original character of the Cuban Revolution. This in turn triggered a process of "Correction of Errors and Negative Trends", which implied a certain dose of self-criticism regarding the mere adaptation of Soviet models since the 1970s, when Cuba had been most deeply integrated into the Socialist system (though always maintaining some degree of autonomy, as we have seen in several examples).[3]

Despite the remarkable results achieved thus far by Cuba's scientific system, its rapid growth of complexity led to a critical attitude towards the low level reached in the application of results, the absence of adequate thematic focus, and the overall lack of an integrated view. As a result, from 1985 on important changes were made in the country's scientific policy. These changes, which aimed at closing the research-application gap, led to the creation of a new type of research centre and of so-called "scientific poles", to the increased utilization of the universities' scientific potential and, finally, to increased innovation linked to the so-called "science and technology forum".

5.5   The Growing Strategic Role of Biotechnology for Achieving Autonomy

Most visible among the transformations of the 1980s was the emergence in Cuba of a high-tech production sector associated with biotechnology, the medical-pharmaceutical industry, and the health care system, in which the process of unlinking the Cuban scientific system from the Soviet one became even more

---

[3]After its early autonomous Third World and non-alignment policy, in 1975 Cuba surprised both Moscow and Washington by returning into the African arena, which Che and his guerrillas had left suddenly ten years before. Castro's decision to intervene militarily in Angola to support the MPLA in the defence of Luanda was so resolute that it took by surprise Brezhnev's intention of a substantial rapprochement with the United States. Without any Soviet logistical support, the fearless Cuban expedition allowed Agostinho Neto to route the assault on the capital and proclaim the independence of Angola. Further Cuban reinforcements led to the historical defeat of the South African army in the six-month long battle at Cuito Cuanavale in 1978–79 and to South African withdrawal from Angola. As Nelson Mandela declared upon his visit to Havana in 1991, the defeat of the racist army at Cuito Cuanavale by a largely black army destroyed the myth of its invincibility and greatly contributed to the subsequent end of apartheid, and was "a turning point for the liberation of our continent and my people".

evident. The development of these sectors is particularly revealing for the argument of the present work, since it clearly illustrates the peculiar features and effectiveness of Cuba's unconventional approach, starting with the acquisition of the most advanced skills in both basic and applied research and arriving at the production of pharmaceuticals and bio-drugs, which was accomplished in a surprisingly short time. This is all the more so as the basic market-focused architecture of today's biotechnological and pharmaceutical industries is a typical product of capital-intensive American industry, whose characteristic attitude to exploiting nature as a commodity has opened the door to the proprietary ownership of living matter (de Sousa Silva 1989, 1995). The Cuban biotechnological system grew independently of this organization and method:

> ...biotechnology in Cuba was driven by public health demand.... In place of an economic imperative, therefore, were the social demands that the Cuban scientists apply to the product of their research as soon as possible.... their model of fast science was... very much against the global trend of the time (Reid-Henry 2010, 18–19).

Almost paradoxically, the outstanding and original progress made by Cuba in the field of biotechnology was also a side effect of US policy, which caused Castro's decision to develop this field in order to overcome Cuba's dependence on pharmaceuticals from other countries (one could recognize in this decision a far echo of Martí's call for overcoming subalternity and achieving full independence).

In early 1980, when the field of biotechnology was still in its infancy in the world, the Cuban government launched a series of programs aimed at applying new biotechnologies to the health care sector in order to address domestic health problems. This choice could rely on an efficient and well-distributed health system, medical schools, research institutes, and pharmaceutical and medical products factories. With the intent of bringing together biology centres, groups and specialists, in order to create the bridge between science and the national economy and to promote and coordinate further progress, in 1981 the *Frente Biológico* (Biological Front) was created. It was established under the auspices of the Cuban Council of State, underscoring a constant feature of Cuban biotechnology: i.e. the government's long-term vision in supporting a sector not likely to produce short-term returns for the investments made. This vision turned out to be a key to Cuban success in the field of health biotechnology (Thorsteinsdóttir et al. 2004a). The high level and the good organization of the system of education played an important role in the success of this new field of research. Centralized control of this system in fact allowed the planning of human resource development, making sure that educational possibilities met national needs, for instance by limiting entry to other disciplines or specialities. In 1981 elite high-standard secondary schools for the sciences were created, where entrance for both students and faculty was highly competitive. During and after university training the best students were selected for internship at research institutions (we have already mentioned the insistence on strong preparation in basic fields, like physics).[4]

The first scientists engaged in the biotechnological sector were physicians. The initial plan foresaw the rapid acquisition of international know-how and technologies and their translation into products of use. In that period, some larger Latin American countries also set up programs for the development of biotechnology (Peritore and Galve-Peritore 1995), but it was Cuba that achieved this goal in a surprisingly short time. One of the explanations for this difference was probably, significantly, the lack of trained personnel in the other Latin American countries (Reid-Henry 2010, 29 and 176 n). By contrast, in Cuba there were an adequate number of biologists who had been trained in the early intensive courses of the 1970s (Sect. 4.6) and had subsequently received specialized training in Italy, France and other countries.

5.6 Entering Modern Biotechnology from Its Beginnings: Obtaining Interferon for the Country's Own Needs

At the turn of 1980s a group of Cuban physicians took an interest in the recently developed interferon[5] for treating cancer. In fact, after endemic diseases had been eradicated in Cuba after effective vaccination programs,[6] cancer and heart diseases took priority as public health issues (Cooper et al. 2006; Health in the Americas 2012, 244). From 1980 on, these physicians established contacts with leading US and Finnish scientists who had studied and produced this protein (Feinsilver 1993a; Reid-Henry 2010, 13–19).

A decade later, Cuba was the pharmacy of the Soviet bloc and Third World (Starr 2012).

First of all, these Cuban physicians visited the American oncologist Randolph Lee Clark (1906–1994), director of the MD Anderson Cancer Hospital in Houston and one of the most eminent American oncologists. Clark in turn visited Cuban health facilities and met Fidel Castro, leaving him convinced that interferon was the right drug to develop. As a consequence of this meeting, two Cuban doctors were hosted in Houston. Clark directed the Cubans to Helsinki, where Kari Cantell had

---

[4]In other systems, for example Italy, biology students take only a one-semester course in physics, in Cuba they do four semesters of physics.

[5]Interferons are proteins made and released by host cells. They trigger the protective defences of the immune system aimed at the eradication of pathogens or tumour cells.

[6]Vaccination in Cuba, which begun as early as in 1962 under the National Immunization Program (NIP), considerably reduced the infectious disease burden contributing to elimination of diseases such as: poliomyelitis (1962), diphtheria (1979), measles (1993), pertussis (1994), and rubella (1995) (Reed and Galindo 2007). It is worth noting that the current vaccine schedule targets all Cuban children for immunization against 13 diseases with 11 vaccines, eight of which are produced by the country's Scientific Pole.

developed a technique for making useful amounts of industry-standard interferon from human blood cells.

The visit of the Cuban scientific delegation in Helsinki and its outcomes occupies an important place in Cantell's memories, and his vivid memory of it is very revealing for us (Cantell 1998, 141–153). At the beginning of 1981, Cantell was officially asked by the Cuban ambassador to Finland to accept a group of visitors from Cuba to learn how to make and purify leukocyte interferon. Cantell was pressed by the great amount of publicity being given to the topic, and consequently he

> asked that the numbers in the "delegation" should be kept to a minimum, and the duration of the visit to a week. At the time, I was fairly sure that the visit would be a complete waste of my time, but I did not want to abandon our open doors policy.On Monday March 30th a group of six Cubans, virologists, immunologists and biochemists, headed by Manuel Limonta, a specialist in internal medicine, came to my laboratory. They were all tired and jet-lagged after their long journey, but they set to work without delay to try to understand our process; they followed our procedures in detail and took copious notes.The team returned to Cuba, and at the beginning of May, I had a letter from Limonta to tell me that their interferon laboratory was nearly finished (Cantell 1998, 142).

Cantell was officially invited to visit the laboratory, but he was not keen to go and so he sent his co-worker, Sinikka Hirvonen.

> When Sinikka returned, her story astonished me. Interferon production was in full swing in Cuba in a laboratory converted from a former luxury house in a suburb of Havana [7] (Cantell 1998, 142).

Fidel Castro was very much involved with the project and visited the researchers every day.

In only a few months the Cuban scientists had observed Cantell's technique, purchased the necessary equipment and materials, reproduced the process in a house equipped as a laboratory (then called "Sinikka's house": Cantell 1998, 143) and stabilized its production.

For the first time Cuba entered an industrial sector at the very moment it was being born globally. But once again, as in the case of superconductivity, this observation is only a formal analogy, and says little or nothing about the peculiar aspects of the Cuban approach. In fact, apart from this coincidence (or "self-narration", in Reid-Henry's terms, p. 20), the specific conditions, goals and mechanisms that underlay scientific development in Cuba tell a special story.

> Cuban biotechnology neither boomed nor coincided.... In place of a boom something rather more modest and considerably more interesting took place: biotechnology developed there for different reasons and in a different way than had been the case elsewhere, certainly in rich Western countries. This was to have profound effects both on the science itself and on the development of that science within the revolutionary machine (Reid-Henry 2010, 21–22).

---

[7]Curiously enough, Clark had also initially begun working in 1946 in Houston with 22 employees in a carriage-house of a donated family estate, equipped with research laboratories for biochemistry and biology, then moved to the estate grounds, and converted to a clinic.

Almost immediately, another characteristic feature of Cuba's practice came to the fore—as early as June 1981 Cuban doctors begun to use interferon in medical practice during a virulent epidemic of haemorrhagic dengue fever.[8] Cantell's vivid account is again expressive.

> Furthermore, [as Sinnika returned] clinical studies with interferon had already begun in a virus infection called dengue.... the possibility of trying interferon therapy in dengue interested me greatly. Quite unexpectedly, I soon got some personal practical experience in this connection. Sinikka had been bitten by mosquitoes in Havana, and, soon after her return, she became severely ill with a disease which according to the textbook was a classical case of dengue.... two days later she came out in a blotchy rash. I gave her interferon injections and she soon recovered—whether the interferon played any part in this, or whether she should have recovered equally quickly without it, is impossible to say (Cantell 1998, 142–43; see also Reid-Henry 2010, 16–18).

We will see further on that the close link between research and clinical testing and application will remain a peculiar feature of the Cuban biomedical system.

5.7 The Leap Towards Genetic Engineering

On the basis of the achievement in producing and using interferon, in response to the urgent need to produce greater quantities, the decision was taken to create the *Centro de Investigaciones Biológicas* (CIB, Centre for Biological Investigation), which was built in only six months. Cantell was invited to the inauguration to cut the blue ribbon, and he "chattered with Castro" (Cantell 1998, 143). He gave two lectures on interferon in the Academy of Science.

The initial work on purification of interferon done together with Cantell (who retired a few years later) was complemented by a parallel project, the attempt to clone interferon, a result that few others had obtained. The orientation towards genetic engineering was not driven in Cuba by the logic dominating in Western industry, or by the search for cutting edge scientific results, but because it did the job, responding to national needs. Molecular biologist Luis Herrera (the first specialist trained in the early 1970s Italian courses, Sect. 4.6) from CNIC was chosen for this job. He was immediately sent to the Pasteur Institute in Paris, and then put in charge of the group created to obtain recombinant interferon. They succeeded in this undertaking as early as 1984, developing a whole new approach from Cantell's

---

[8]Dengue fever is a disease caused by any one of four related viruses transmitted to humans by mosquitoes. It can cause severe flu-like symptoms and in severe cases can be fatal. Dengue has emerged as a worldwide problem only since the 1950s. With more than one-third of the world's population living in areas at risk for infection, the dengue virus is a leading cause of illness and death in the tropics and subtropics. As many as 400 million people are infected yearly. There is no vaccine or medication that protects against dengue fever.

technique: theirs was a second-generation, recombinant interferon (Reid-Henry 2010, 46–47).

> By 1986 Cuba was "the second-largest producer of natural human leukocyte interferon, after Finland"[9] … According to California-based Genentech researcher Patrik Gray, "the Cuban production system is pretty much like that of other groups using yeast alpha-factor, but what is different is that they're using it to produce interferon for clinical purposes" … Interferon was chosen as a model to develop genetic engineering and biotechnology techniques because it was then thought to be a potential wonder drug, particularly in cancer treatment and as an antiviral medicine, and, more important, because it served as a model for the development of advanced molecular biology skills… A US biotechnology industry analyst substantiated the Cuban approach when he suggested in 1990 that "alpha interferon almost serves as a paradigm for all of these biological response modifiers…" which are at the forefront of biotechnology research (Feinsilver 1995, 101).

In fact, even though interferon has not become the magic wand in treating cancer as originally hoped, it was important in boosting learning, the confidence of Cubans in their biotechnology, and the start-up of autonomous projects. Between 1982 and 1986 the development of molecular biology and genetic engineering at the CIB represented the first step leading to their own innovations and development of knowledge.

5.8 Ends Above Means: Differentiating from Mainstream Western Biotechnology

An instrumental breakthrough occurred between 1982 and 1984, which further highlighted the originality of Cuba's approach and goals. In 1981 the UN Industrial Development Organization (UNIDO) announced a competition for an international centre to promote research and development in biotechnology in the Third World. Cuba applied, among over fifteen other countries. But when the final decision had to be taken, Cuban scientists realized that

> national needs would never be met within a framework designed and operated by the advanced industrial nations (Reid-Henry 2010, 45).

As a consequence, in 1983 the decision was taken to autonomously construct a new institution devoted to the development and application of genetic engineering. Thus, the *Centro de Ingeniería Genética y Biotecnología* (CIGB centre for Genetic Engineer and Biotechnology) was inaugurated in 1986 (Reid-Henry 2010, 53–57). Randolph Lee Clark attended the opening of the centre.

With a cost of roughly $25–26 million (in the US it might have cost 10 times more: Reid-Henry 2010, 53) and a further investment of roughly $100 million to supply it with the most advanced equipment and facilities for research in molecular

---

[9]"Cuban-made interferon reaches out for world markets", *Newswatch*, March 17, 1986: 3.

genetics and genetic engineering, the CIGB was to become the largest scientific centre in Cuba: a low–cost, high return enterprise.

> The quality of Cuban equipment and research facility at CIGB was of Japanese or West European level, but due to the US trade embargo it was very costly, since procurement of materials and equipment in Japan and Europe means high transportation costs and delivery delays. Consequently, Cuban scientists had to learn to produce their own restriction enzymes, make tissue cultures, establish virus collections, as well as to develop and manufacture equipment to do electrophoresis and gas chromatography (Feinsilver 1995, 103).

With a concentration of hundreds of researchers, the CIGB was divided into small groups that covered practically the whole spectrum of topics in the field. As its mission, the centre assumed the responsibility of directly contributing to the social-economic development of Cuba. Its fields of research go from human health, to agricultural and aquaculture production, industry and the environment. Through research the CIGB generates knowledge for the development of new products, services and marketing based on a quality system. The Centre's experimental social and spatial innovations allowed it to originate new modes of scientific practice and to become an engine for economic development. The CIGB's activities encompassed: production of proteins and hormones; development of vaccines and pharmaceutical products; research on genetic engineering of microorganisms and plant and animal cells; production of enzymes; development and production of diagnostics. The CIGB's functions ranged from R&D to production, and later also the commercialization of its products and diagnostic equipment, through its own commercial agency, which had the status of a limited company.

The full-cycle conception was an explicit strategy in Cuban biotechnology, made easier by centralized state control, as many commentators have remarked (Elderhost 1994; Thorsteinsdóttir et al. 2004a; Reid-Henry 2010).

The development of a national capacity in biotechnology was seen as a strategy to increase sovereignty and independence from the transnational companies of the industrialized countries, especially in the medical sector, principles that Cuba has always advocated within the movement of nonaligned developing countries. In the 1980s Cuba was already acting in the major markets in Eastern Europe and the former Soviet Union, and one of its first attempts was to promote technology transfer within the COMECON, an alliance of countries that did not recognize Western intellectual property law. But at the same time, and coherently with its entire position, Cuba tried to increase scientific relations with the West (Reid-Henry 2010, 55–56; de la Fuente 2001), a strategy that proved to be all the more important when the former approach was soon after made impossible by the unforeseen collapse of the Soviet Union (Chap. 6).

5.9   The First Great Achievements and Further Implications of a Need-Driven Approach

In the following years the range of interests and activities of Cuban biotechnologists considerably widened, in connection with and under the stimuli of the public health system. It now included several recombinant molecules like colony-stimulating factors, interleukins, monoclonal antibodies,[10] and vaccines.

The early approach to cancer, developed in the National Oncology and Radiobiology Institute (INOR) by a group of theoretical biophysicists who modelled antigens and cellular receptors related to some diseases and cancers affecting the Cuban population, laid an important foundation for the subsequent approach to this illness (Chap. 6). In order to more adequately carry on the cancer research started at the INOR, at the turn of the 1980s a completely new, very modern and well-equipped institute was created, the *Centro de Inmunología Molecular* (CIM, centre for Molecular Immunology), after a specific proposal by Fidel Castro (see later).

In 1982 an original, advanced diagnostic tool was created, the Ultramicroanalytic System (SUMA) for diagnosing human diseases such as HIV, hepatitis B, herpes simplex, Chagas, dengue, leprosy, and congenital defects. This and other diagnosis tests are exported to Brazil, Spain, Colombia and other countries.

As early as 1982 Cubans were producing the first monoclonal antibodies, developed in 1975 by American researchers G. Kohler and C. Milstein, who were awarded the Nobel Prize for this discovery in 1984. In Cuba clinical tests with monoclonal antibodies started in 1984, and by 1989 they were clinically applied as anti-rejection drugs, while other possible applications were being examined, since they offered the possibility of delivering target therapies (see Chap. 6).

> A Monoclonal Antibody Program was launched at the National Institute of Oncology and Radiobiology (INOR) in Cuba in 1985…. These monoclonal antibodies are especially useful in immunohistochemistry for the study of human tumours, leukaemia and lymphomas…. The use of anti T-cell monoclonal antibodies for the prevention of rejection crisis in patients receiving organ transplantation and for the treatment of patients with cutaneous T-cell lymphoma is still mostly a matter for research (Lage 1990).

As we shall see in the Chap. 6),

> The availability of monoclonal antibodies in Cuba has facilitated development and application of innovative techniques (immunoscintigraphy and radioimmunotherapy) for cancer diagnosis and treatment (Peña et al. 2014).

---

[10]A *colony-stimulating factor* is one of a number of substances secreted by the bone marrow, that cause stem cells to proliferate and differentiate, forming colonies of specific blood cells. The function of the immune system depends in large part on *interleukins*. This is a generic term for a group of secreted proteins from white blood cells that stimulates their activity against infection and may be used to combat some forms of cancer. A *monoclonal antibody* is an antibody produced by a single clone of cells ('hybridoma') grown in culture, capable of proliferating indefinitely to produce unlimited quantities of identical antibodies (in contrast to polyclonal antibodies that are made from several different immune cells).

Great advances were also made with vaccines. One of the most relevant results of the Cuban biomedical sector was the achievement in 1986 of a vaccine against meningococcus B, which is the only proven effective one in the world. Indeed, it received international certification and won the World Intellectual Property Organization's gold medal for innovation[11] (Reid-Henry 2010, 64–66). It was obtained in a purpose-established laboratory in the Ministry of Public Health, in order to fight meningococcical meningitis, which in the early 1980s was the main infectious disease in Cuba (an epidemic had broken out in 1976). During the year 1989–90, over 96 % of Cuban population groups at risk were vaccinated, with the usual close correlation between research and clinical testing.[12] As soon as the vaccine began to be used on a mass scale in Cuba, requests arrived from Brazil, Argentina, Colombia and other countries. In 1989 the laboratory was transformed into the *Centro Nacional de Vacuna Antimeningocócica* (National Centre for Anti-meningococcal Vaccine), incorporated in 1991 in the new *Instituto Finlay*, a centre comparable in size to the CIGB (20,000 square meters of installations: Reid-Henry 2010, 65) but specialized in R&D for serums and vaccines.

In the meantime, since the technological bases for realizing the first preventive program for genetic diseases and congenital defects in Cuba already existed, the first national program for the diagnosis and prevention of genetic diseases was established on a proposal by Fidel Castro. This program focused on the health priorities of the Ministry of Public Health (Lantigua Cruz and González Lucas 2009).

It is worth adding some comments on these developments, which we shall reconsider in more detail in Chap. 6. On the one hand, as has been remarked,

> Preventive vaccines... are particularly attractive because in several cases recombinant vaccines are available where the technology for producing them is relatively simple and can easily be reproduced. In general, they are a much more efficient and cost-effective way of dealing with infectious diseases than drugs, and they can provide long-term immunity. They are very relevant to the health agenda of most developing countries with large, poor populations

(Thorsteinsdóttir et al. 2004b, 50). In fact, the close ties of Cuban biotechnology with the national health care system have more general implications.

> Cuban research also prioritizes developing affordable vaccines for diseases affecting poor populations, such as typhoid fever and cholera: a fundamentally needs-driven, rather than market-driven approach. This can be contrasted with transnational pharmaceutical companies, which have come under increasing criticism for placing market interests before global health solutions, resulting in investment of 90 % of R&D dollars worldwide in developing treatments for diseases affecting the 10 % of the world's population that can afford the results

(Evenson 2007).

---

[11]"U.S. finally will let SmithKline market Cuban meningitis vaccine", *Wall Street Journal*, July 23, 1999.

[12]Reid-Henry (p. 66–67) discusses an interesting comparison of the practices of research and clinical testing in the case of this vaccine adopted by Cuba and by Norway.

As a business development manager at the CIGB has said: "The success is not sales, it's the impact on society. We are chasing problems, not profits" (Manuel Raíces Pérez-Castañeda, reported by Giles 2005).

5.10   A Sound Network of International Relations

The attainment of a high level of scientific research at an international standard and of autonomous development does not imply that research can proceed in isolation particularly in modern science. Autonomy does not mean autarchy, but on the contrary the capability of interchanging information and progress with the international scientific community at the highest level and on an equal footing. Ideas move with people and circulate with interchange. We have discussed how, after the achievement of a critical mass of well-trained professionals, Cuban scientists could easily integrate in every international scientific milieu. During the 1980s Cuban scientists extended and reinforced their interchanges and training in many foreign countries, including France, Germany, Japan, the United States, Finland, Belgium, the Netherlands, the United Kingdom, Sweden, Switzerland, Italy, Argentina, Brazil, Mexico, the USSR and Czechoslovakia.

> There was a surge of Cubans obtaining foreign training around 1980 when biotechnology was a new field. Many studied at the pioneering life science institutions in Western Europe and the United States, including the Curie Institute (Paris), the Pasteur Institute (Paris), Heidelberg University (Heidelberg, Germany) and Harvard University (Cambridge, MA, US). This helped shaping the thinking of some of Cuba's current leaders of health biotechnology research. As the sector has advanced, the knowledge flows have become more varied and are increasingly both to and from Cuba

(Thorsteinsdóttir et al. 2004a, 23; Bravo 1998). This is all the more important since, due to the predominantly mission-oriented (though not purely applied) framework of research adopted in Cuba, studying abroad served to integrate training and updating in basic research, a sector that is also considered very important but one that is too expensive to develop fully without extensive funding.

By 1989 in Cuba there were 41,784 science workers in all fields (one for every 251.3 inhabitants), of whom only 120 had doctorates and 2,192 were doctoral candidates in the European sense.

Incidentally, the initial choice of developing a solid, multi-purpose physics sector proved correct. Combined with the particular facility with which Cuban scientists could shift from one specialty to another, and the broad basic preparation and versatility of Cuban physicists, this choice fostered cross-disciplinary links and reciprocal understanding, while at the same time avoiding excessive specialization. From the late 1980s the rapid growth of biotechnology and of the medico-pharmaceutical industry attracted dozens of newly graduated physicists to the new centres (Baracca et al. 2014b, 208–209). Most of these young scientists continued their training in molecular biology and became specialists in this field, and some went on to achieve great distinction. Moreover, other institutions related to physics directed their efforts totally or partially towards biomedical applications, their most frequent activities being in the realm of equipment manufacture and software design.

5.11   An Integrated Biomedical Network

A further leap forward of the Cuban biomedical complex was made at the end of the 1980s. In 1991 all the centres around the CIGB were grouped into the *Polo Científico del Oeste* (Western Scientific Pole, of Havana), which besides the CIGB included the already mentioned *Instituto de Medicina Tropical "Pedro Kourí"* (IPK) and the following centres: *Centro Nacional de Producción de Animales de Laboratorio* (CENPLAB 1982, National Centre for Production of Laboratory Animals), *Centro de Inmunoensayo* (1987, Centre for Immunoassay), *Centro Químico-Farmacéutico* (CQF 1989, Chemical-Pharmaceutical Centre), *Centro de Neurociencias de Cuba* (CNC 1990, Cuban Centre of Neurosciences), *Centro de Inmunología Molecular* (CIM 1991, Centre for Molecular Immunology), the *nuevo Instituto Finlay* (1991, new Finlay Institute),[13] and later the *Centro de Biopreparados* (1997, Centre for Bio-drugs).

Note that these steps forward were made while connections with and support from the Soviet Union were deteriorating, culminating with the final collapse of the USSR (see next Chapter). This confirms that reliance on science was one of the main pillars of Cuban strategy. And these decisions, taken under extremely adverse conditions, reinforced the original features of the Cuban scientific system.

The tight interconnection of the multiple centres grouped together in the Scientific Pole, its style of work and its motivations fostered an original technical culture and forms of scientific activity, with new connections between mission-oriented research and basic science, "a particular sort of epistemic space", of "creation of innovation" (Reid-Henry 2010, 85–87). An innovative spatial experimentation made possible new ways of thinking about scientific practice itself. A peculiar "practical epistemology", an alternative mode of scientific behaviour, shaped Cuban biotechnology, which had grown out of the happy interpenetration-synergy between a space of creativity and innovation and an epistemic approach, the integration of what people are doing with the context in which they work.

Not only close scientific cooperation, but also socialization and community have been promoted. Researchers are given responsibility at a relatively early stage, and there is strong emphasis given to ethics regarding one's own conduct, mobility and cross-disciplinary links.

The particular form that Cuban biotechnology took allowed it to achieve substantial scientific advances without the need for the huge resources invested in Western biotechnology.

---

[13]In 1927 a former *Instituto Finlay* had been created in Havana.

> By the late 1980s, Castro's vision that the future of the nation had necessarily to be a future
> of men of science had been realized in a form that no one—and perhaps last of all himself
> —would have predicted

(Reid-Henry 2010, 87).

Moreover, what had happened with the direct experimentation of interferon in health care in 1981 was not an isolated case, but a specific feature of Cuban biotechnology, which was driven by public health demand and the close integration of the health system with the system of research and pharmaceutical production (the "Cuban fast-track approach to developing new drugs": Reid-Henry 2010, 17–18).

> The health sector in Cuba is heavily involved in clinical trials, and many clinicians are
> intimately involved in the whole research process. For example, the research institute,
> Pedro Kourí Institute in Tropical Diseases (Havana), incorporates a hospital and plays a
> leading role in the evaluation and clinical trials for all Cuban vaccines. The tight integration
> of research with the health system and the Cuban health biotechnology sector encourages
> the creation of innovative products. A research director at a public research institute said,
> "We have feedback from the clinical trials to the lab. This is not a linear process. The cycle
> is a good ground for innovative thinking. It has definitely improved our products." The
> health sector is therefore not only a recipient of innovation but actively promotes it as well
> (Thorsteinsdóttir et al. 2004a, 22).

It was also decided to extend the development of modern biotechnology to the rest of the country, creating the following centres: two more CIGB in the provinces of Camaguey (1989) and Sancti Spiritus (1990), and two *Centro de Biotecnología* in Ciego de Ávila (1991) and Villa Clara (1992).

References

Arés Muzio O, Altshuler E (2014) Superconductivity in Cuba: reaching the frontline (Baracca A, Renn J, Wendt H, eds) 2014:301–306

Baracca A, Fajer Avila VL, Rodríguez Castellanos C (2014a) A comprehensive study of the development of physics in Cuba from 1959 (Baracca A, Renn J, Wendt H, eds) 2014:115–234

Baracca A, Renn J, Wendt H (eds) (2014b) The history of physics in Cuba. Springer, Berlin

Bravo E (1998) Development within under-development: new trends in Cuban medicine. Editorial José Martí/ElfosScientiae, Havana

Cantell K (1998) The story of interferon: the ups and downs in the life of a scientist, World Scientific

Cooper RS, Kennelly JF, Orduñez-García P (2006) Health in Cuba. Int J Epidemiol 35(4):817–824

de la Fuente J (2001) Wine into vinegar: the fall of Cuba's biotechnology. Nat Biotechnol 19:905–907

Elderhost M (1994) Will Cuba's biotechnology capacity survive the socio-economic crisis? Biotecnol Dev Monit. 11:13-22

Evenson D (2007) Cuba's Biotechnology Revolution. MEDICC Rev 9(1):8–10

Feinsilver JM (1993) Healing the masses. Cuban health politics at home and abroad. University of California Press, Berkely, CA

Feinsilver JM (1995) Cuban biotechnology: the strategic success and commercial limits of a first world approach to development. In Peritore and Galve-Peritore

Giles J (2005) Cuban Science: ¿Vive la revolution?. Nature, 436:322–324

Health in the Americas. 2012 Edition. Country Volume *Cuba*. Pan American Health Organization

Lage A (1990) The monoclonal antibody program of the cuban institute of oncology and radiobiology. Eksp. Onkol. 12(6):26–30, 64

Lantigua Cruz A, González Lucas N (2009) Development of medical genetics in Cuba: thirty nine years of experience in the formation of human resources. Rev Cubana Genet Comunit (internet). 3(2):3–23. http://bvs.sld.cu/revistas/rcgc/v3n2_3/ Accessed March 15 2016

Peña Y, Perera A, Batista JF (2014) Immunoscintigraphy and radioimmunotherapy in Cuba: experiences with labeled monoclonal antibodies for cancer diagnosis and treatment (1993–2013). MEDICC Rev 16(3–4):55–60

Peritore NP, Galve-Peritore AK (eds) (1995) Biotechnology in Latin America: politics, impacts and risks. Scholarly Resources, Wilmington, DE

Reed G, Galindo MA (2007) Cuba's national immunization program. MEDICC Rev 9(1):5–7

Reid-Henry S (2010) The Cuban cure: reason and resistance in global science. University of Chicago Press, Chicago

de Sousa Silva J (1989) Science and the changing nature of the struggle over plant genetic resources: from plant hunters to plant crafters. PhD Thesis. The University of Kentucky, Lexington. United States

de Sousa Silva J (1995) Plant intellectual property rights: the rise of nature as a commodity. In Peritore ad Galve-Peritore 1995:57–68

Starr D (2012) The Cuban biotech revolution. http://www.wired.com/2004/12/cuba/ Accessed March 15 2016

Thorsteinsdóttir H, Sáenz TV, Quach U, Daar AS, Singer PA (2004a) Cuba. Innovation through synergy. Nat Biotechnol. 22 (Supplement) December: 19–24

Thorsteinsdóttir H, Quach U, Daar AS, Singer PA. (2004b) Conclusions: promoting biotechnology innovation in developing countries. Nat Biotechnol. 22 (Supplement). December: 48–52

Chapter 6
Decisive Results … and New Challenges

> *"The Cuban internationalists have made a contribution to
> African independence, freedom and justice, unparalleled for its
> principled and selfless character...Cubans came to our region
> as doctors, teachers, soldiers, agricultural experts, but never as
> colonizers. They have shared the same trenches with us in the
> struggle against colonialism, underdevelopment, and
> apartheid."* [Nelson Mandela's address at the opening of the
> Southern African Cuba Solidarity Conference in Johannesburg
> in October 1995, The Nelson Mandela Foundation] (https://
> www.nelsonmandela.org/omalley/cis/omalley/OMalleyWeb/
> 03lv02424/04lv02730/05lv03005/06lv03006/07lv03105/
> 08lv03112.htm Last access March 15, 2016.)

Abstract At the end of the 1980s Cuba was hard hit by the unexpected sudden
dissolution of the Soviet Union and the socialist market. The substantial *resilience*
of the Cuban scientific system in the face of this event, despite unavoidable set-
backs, confirmed its achieved autonomy. What seems remarkable is that even in this
extremely critical economic situation the Cuban government once again confirmed
its support for science as a strategic choice. And up to now this has proved a trump
card. True, several scientific sectors had to be downsized or redirected, but for
health biotechnology unconditional support was confirmed and even reinforced.
Incorporation in the unprotected global market has raised new challenges. These are
not only of an economic nature but also involve, for instance, rights to intellectual
property, which in Cuba remains social. Today biotechnology represents Cuba's
third source of hard currency, with healthcare products and services representing a
major island's export potential. In the end we can say that the success of a small
country with limited resources in a typically American-dominated, capital-intensive
field like biotechnology is probably Cuba's most remarkable and enduring
achievement.

Keywords Healthcare · Biotechnology · 'Periodo especial' · Embargo towards
Cuba · BioCuba Farma · Cancer immunotherapy

© The Author(s) 2016
A. Baracca and R. Franconi, *Subalternity vs. Hegemony, Cuba's Outstanding
Achievements in Science and Biotechnology, 1959–2014*, SpringerBriefs
in History of Science and Technology, DOI 10.1007/978-3-319-40609-1_6

6.1   A "Disaster Proof" Scientific System

The overcoming of the condition of subalternity and the attainment of full autonomy, even in a specific sector, is never a final result, guaranteed once and forever.
This is especially true for a developing country, and even more when it is small and
in an extremely critical geopolitical situation. Other, bigger and richer, countries
have undergone mixed fortunes in their history, power and decline. In the contemporary world, moreover, the pace of events runs increasingly fast and can
overwhelm even big powers. This was precisely what happened, absolutely
unexpectedly, at the end of the 1980s with the sudden downfall of the Socialist
block and the dissolution of the Soviet Union, a country incomparably bigger and
stronger than Cuba.

Over and beyond its brilliant achievements, it was just these dramatic events that
proved that the Cuban scientific system had indeed overcome the condition of
subalternity and the attainment of full autonomy.

The consequences for Cuba were terrible. The overall economy was reduced by
almost 50 % between 1989 and 1992, and the gross domestic product decreased by
30 % (ONE 1996). Traditional Cuban exports of sugar, nickel, tobacco and seafood
sharply dropped off. However, this is not the place to enter into these aspects, which
are discussed in detail in numerous publications.

Most analysts predicted the downfall of the Cuban economy and regime. Yet,
after a quarter of a century from the downfall of the socialist system, Cuba is still
there. The American blockade was even tightened. Cuba (for better or for worse)
stands firm on its principles albeit the conquests of the Revolution may be in part
undermined or transformed by economic difficulties. Moreover, for a couple of
decades the isle has played a renewed guiding role in the trend of transformation
and integration of the Latin American and the Caribbean continents, which followed the end of dictatorial, military or reactionary regimes. Let us simply recall
that just in December 2011 the CELAC (Community of Latin American and
Caribbean States) was created, which joins all the states of the Continent, with the
exception of the United States and Canada: it was a sort of "alternative" to the
Organization of American States, from which Cuba had been expelled in 1962: it
was no coincidence that Cuba got the CELAC first presidency.

As far as our subject is concerned, not only did the Cuban scientific system
basically resist the impact, but the strategic choice of creating an advanced scientific
system was also stubbornly upheld. And, we could say, the effectiveness of this
choice was confirmed.

Obviously, scientific activities were profoundly affected. For instance, in physics
experimental activities met great difficulties due to the shortage of spare parts of
equipment of Soviet origin. This shortage was worsened by the blockade.

Irreplaceable scientific information and updating had to be cut, due to the prohibitive subscription fees of international journals.[1]

In physics and in other scientific sectors several activities declined, or even had to be closed down or redirected (Baracca et al. 2014). For instance, the project of the nuclear power plant in Juraguá was abandoned after the failure to find alternative international partners. Nevertheless, on the whole the Cuban scientific system resisted the impact of the fall of the USSR, showing the toughness, cohesion and level of autonomy and self-sufficiency it had reached over the previous three decades. Once more, in this critical circumstance the ability of Cuban scientists (not to say of Cubans in general) to make a virtue of necessity and to exploit their own resourcefulness proved successful.

The widespread international collaborations established by the Cuban scientific community throughout the previous decades proved all the more useful in the new situation, since they provided alternative options and invaluable opportunities (see Baracca et al. 2014, Part Three). The physicists intensified their relations and exchanges with Spain (although with more difficulty after the 2008 financial crisis), Mexico, Brazil and other countries, besides the International Centre of Theoretical Physics in Trieste.

6.2 Meeting a New Challenge

But the problem was not merely to resist in such dramatic circumstances. To this end, in 1990 a *Periodo Especial* was proclaimed ("Special Period in time of peace"), in which the Cuban population would have to cope with severe difficulties in the following years. As a consequence, the general nutrition of the population suffered, per capita caloric intake being reduced by 24 % (FAO 2003).←

On its part, the United States tightened the economic blockade with even more restrictive laws (Torricelli Act 1992, and Helms-Burton Act 1996).[2] Such a stranglehold on Cuba's economy worsened shortages of food and medicine (Kirkpatrick 1996).

---

[1]An expedient observed by one of us (A.B.) in Cuba around the mid-1990s, and typical of Cuban initiative, was to resort to much less expensive personal subscriptions by students to some journals when funds were available. These were later transferred to the faculty the following year.

[2]The "Cuban Democracy Act" was a bill presented by US Congressman Robert Torricelli and passed in 1992 which prohibited foreign-based subsidiaries of US companies from trading with Cuba, travel to Cuba by US citizens, and family remittances to Cuba. The deeply controversial "Cuban Liberty and Democratic Solidarity Act" of 1996 (commonly known as Helms–Burton Act) is a United States federal law which strengthens the US embargo against Cuba, extending the prohibition of trade with Cuba to companies doing business with it and to companies that use property Cuba had nationalized in 1958 from American companies. The US Government prosecuted Merck, the largest pharmaceutical firm in the US, for an exchange of scientific information with Cuba.

In the face of this situation, Castro chose to prioritize three economic sectors for investments, in order to keep the Cuban economy and the Revolution afloat: tourism, food production, and biotechnology/medical exports. In a word: once and again, the renewed challenge was met by relying on science as a driving force of economic recovery and reaffirmation of the autonomy and independence of the country. The Fourth Congress of the Communist Party of Cuba, held in 1991, made this choice imperative.

> Faced with economic calamity, Castro did something remarkable: he poured hundreds of millions of dollars into pharmaceuticals» (Starr 2012).

The development of biotechnology aimed at import substitution, the creation of new export products and supporting the national program of food production. Obviously the situation was radically different from thirty years before, both domestically and internationally. Even the most favourable analysts wondered if this choice would prove successful:

> Can biotechnology save the revolution? (Feinsilver 1993).

If in 1960 the starting conditions may have seemed prohibitive, now the troubles and the uncertainties were no smaller.

Could Cuba find sufficient markets and ways to commercialize its products in the globalized capitalistic market, controlled by transnational pharmaceutical companies with patents and well-organized markets and distribution networks, besides enormous capital and finances? In the past the primary goals had been import substitution and commercialization in the Socialist market, but export had become increasingly critical when the government tried to replace lost aid and trade. The effect of the economic blockade on these spheres was particularly damaging. Intellectual property rights represented one more obstacle. Cuba was seeking new, more competitive processes of making already patented products, which would allow it to apply for new patents. But, without past experience or commercial organization, would Cuba succeed in carving out a niche for itself in the world market? For almost three decades the Cuban economy had adapted to guaranteed and protected markets for standardized products, lacking the level of efficiency and competitiveness of capitalist markets, which have specific commodity-related rules of marketable quality.

And, too, would the priority of commercialization, coupled with the limitation of resources, leave space for the investments in basic research that is necessary to make real breakthroughs in the field? And would Cuban society hold out against the increasing divergence between an advanced and privileged technological system and the worsening living and economic conditions, along with the unavoidable deterioration in social services?

Once more, and in very critical conditions, Cuba was moving on the border of the continually growing technical gap between the poor and the rich countries, between subalternity and hegemony.

6.3 Further Impulse to the Cuban Scientific System

Once more the Cuban government's courageous choice seems to have been a trump card. Despite the extremely difficult circumstances created by the fall of the Soviet Union and of the worsening economic embargo, activities in the biomedical field were confirmed and even reinforced. New centres were created (as listed in the previous chapter). These were mainly full-cycle research groups. From 1990 to 1996, the Cuban government invested around 1 billion US dollars in the Western Havana Bio-Cluster, the foremost Cuban biomedical complex, which comprises 52 major research, education, health, and economic institutions devoted to the biotechnology field. Moreover, in 1993 the Centre for Biophysics and Medical Physics was created at the *Universidad de Oriente* (Eastern University), dedicated to R&D in the area of magnetic resonance and its biomedical applications (Cabal Mirabal 2014). As of 1992, there were 53 research centres in biotechnology in Cuba (Feinsilver 1995, 102).

> According to Elena Siméon, the director of the *Academía de Ciencias de Cuba* (ACC), approximately 8,000 people worked in scientific research relating to biotechnology in 200 research institutes in 1993. In the period 1988-1992, more than US$300 million was invested in medical and pharmaceutical industrial biotechnology (Elderhost 1994).

As is natural, the difficult economic situation made it impossible to provide equal, or even comparable, support to all scientific branches. On this point, Giles' report presents the crude reality of the situation in 2005, insisting in particular on existing contradictions (Giles 2005). Nevertheless, even this author had to recognize that:

> Given the hardships suffered by researchers outside the charmed circle of priority applied research projects, it is surprising not to hear more complaints from Cuban researchers. Government control may be one factor; open dissent is a risky policy in a non-democratic country. But equally important is an awareness that Cuba has battled against the odds to avoid the chaos and privations suffered by neighbouring countries such as Haiti. Older Cuban scientists, who remember the right-wing dictatorship that preceded Castro, are especially proud of what's been achieved (Giles 2005).

Speaking in general the period of economic scarcity following the Soviet collapse had to face numerous political choices and considerable re-organization. Centralization of decision-making was reinforced, trying to combine it with more open and dynamic planning processes at lower levels. Five-year planning was substituted for yearly planning, and there were even monthly readjustments.

In the area of biotechnology, on the one hand the organization of scientific work in the Scientific Poles was reinforced, in the attempt to increase efficiency, interdisciplinary cooperation and exchange of knowledge, information and instruments, as well as the rapid implementation of results. On the other hand, the directors of the most important agricultural research institutes and the president of the ACC met monthly in the *Frente Biológico* to coordinate research activity and avoid overlapping. Over all of them there was the supervision of the State Council, which the directors of the CIGB and the Finlay Institute and the president of the ACC were members of.

However, the shift from the socialist to the capitalistic market created very great and complex problems of a completely new nature. Apart from the capitalistic features of commercialization of products, quality control, regulation and standardization, what represented the main obstacle now was intellectual property (IP) rights. It is no coincidence that in those same years the developing countries were reluctant to accept the TRIPS agreement (Agreement on Trade-Related Aspects of Intellectual Property Rights), which came into force in 1995. Just a few years before, in 1986, Cuba had strongly opposed the new global IP rights regime in the Uruguay Round of General Agreementon Tariffs and Trade (GATT).

Cuba previously had a socialist form of regulation of IP, which was the expression of public rather than private interests (intellectual property to the author, *versus* commercial property to the state). But now, in order to acquire hard currency, Cuba had no choice but to integrate into the capitalistic system of IP. Consequently, in 1995 it signed on to TRIPS (Sanchelima 2002; Cárdenas 2009; Plahte and Reid-Henry 2013). However, a solution was found in state ownership.

> The patents of the Cuban industry are owned by the government agency, which avoids the problem of mutual blocking. This agency functions as a kind of patent pool, where every firm has the possibility of using complementary knowledge in advancing new products. ... This resembles more an *internal* open source of innovation [coherent with] the notion of cooperation instead of competition (Càrdenas 2009).

In any case, integration into the international system was done while keeping margins of typical Cuban flexibility, resorting to loopholes in the embargo, and negotiating bilateral agreements with specific countries when possible.

In the end, Cuba succeeded in diversifying and promoting its exports. Already in the early 1990s one of its main biotechnological products exported was the meningitis B vaccine (Chap. 5) that since 1986 had achieved immunity of 97 % of the vaccinated children and adults. It was registered in Brazil (for an export of 15 million doses), Uruguay, Bolivia, Paraguay, Nicaragua, and Asiatic, European and African countries, while Argentina was on the way to registering it, Chile had started experimentation, and Colombia had put a potential epidemic under control in 1990–91. Other exported products were the hepatitis B vaccine, manufactured since 1987, the SUMA equipment, the PPG (a Cuban medicine to reduce cholesterol) and interferon (Elderhost 1994).

A decade later, one study concludes that

> Ten of these institutions [of the Western Havana Bio-Cluster] are at the core of the system as they supply economic support to the whole effort with their production capacities and exports. They are performing more than 100 research projects which have generated a product pipeline of more than 60 new products most of which are protected by intellectual property, and more than 500 patents have been filed overseas (López Mola et al. 2006).

The Economist Intelligence Unit estimated that the increase in non-tourism services exports between 2003 and 2005 was around US$1.2 billion for a total of

US$2.4 billion, which put non-tourism services ahead of gross tourism earnings (of US$2.3 billion) in 2005. Most of these were medical services (Feinsilver 2006).

6.4 More Challenging Choices

From the point of view of the present analysis it is once more very revealing that the choices of Cuban biotechnologists, even under the pressure of the impelling economic situation, continued to follow their own original paths compared to the mainstream directions prevailing in world biotechnology. As for previous developments, these choices depended on various peculiar features of the Cuban situation.

In the first place, incentives coming from the health system continued to determine priorities. Cuba had long reached a first-world health profile, stamping out traditional third-world diseases: chronic illnesses prevailed over infectious ones and cancer constituted one of the main challenges, as the second cause of death after heart diseases for over-65s (a significant portion of cancers, approximately 18 %, have an infectious origin, with a higher incidence in developing countries than developed ones.)[3]

In this context, Cuba could rely on two past achievements. On the one hand, Cubans had accumulated great experience and obtained important results in the therapeutic approach and treatment of infectious diseases. On the other hand, as we have seen in the previous chapter, the availability of monoclonal antibody technology in Cuba facilitated development and application of innovative techniques (immuno-scintigraphy and radio-immunotherapy) for cancer diagnosis and treatment (Lage 2009, 2014; Peña et al. 2014). These premises paved the way for a targeted therapeutic approach to cancer and related treatment, i.e. immunotherapy.

In 1984, in fact, at the INOR Augustín Lage (the future director of the CIM) and his collaborators were the first in the world to understand the role of EGF receptors in breast cancer. EGF (epidermal growth factor) is the name of a class of proteins that stimulate cell growth, proliferation and differentiation by binding to its receptors in the cell surface. Continuing work that Lage had initiated in collaboration with a group in Nice, France, Cuban researchers found that in a very high percentage of human breast tumors EGF receptors (EGFR) were overexpressed. Parallel to the so-called "passive immunotherapy" based on the use of monoclonal antibodies targeting EGFR in cancer cells (which led to the humanized monoclonal antibody *nimotuzumab*, nowadays registered for several tumours, used in over 25

---

[3]Infections cause one in six of all cancers worldwide, International Agency for Research on Cancer, 9 May 2012, http://www.iarc.fr/en/media-centre/iarcnews/pdf/TLO-INF-May2012-Eng. pdf (last access March 15, 2016). Jamal A., Centre M. M., DeSantis C., and Ward E. M., Global Patterns of Cancer Incidence and Mortality Rates and Trends, *Cancer Epidemiol Biomarkers Prev.* 19: 1893–907, August 2010, http://cebp.aacrjournals.org/content/19/8/1893.full (last access March 15, 2016).

countries and in clinical trials for ovarian, pancreatic, lung, stomach and uterine cancers) Cuban scientists developed a new approach, using drugs that do not target the tumour itself, but the immune system ("active immunotherapy").

> Immunotherapy is emerging as an alternative, with new monoclonal antibodies, therapeutic vaccines and deeper understanding of fundamental phenomena in the interaction between tumor and immune system (Lage 2014).

In short, at variance with research on EGF pursued in many other places, the idea was to turn a problem into its solution, that is, to employ human EGF (the drug) as an active factor to use to evoke an immune response able to hinder the process of binding EGF to its receptors in cancer cells and/or to kill tumour cells displaying EGFR.

> This approach potentially offered the advantage of requiring just one or two doses, and of being cheaper to develop (Reid-Henry 2010, 99),

a therapeutic approach that was therefore

> the opposite of what the global pharmaceutical industry might have wanted.

For this transition to an active immunological approach to cancer, Cubans were better positioned than other countries, thanks to the long experience of the vaccinologists at the Finlay Institute working in meningitis B (previous Chapter)

> An active immunotherapy approach was thus rather more conceivable in Cuba than elsewhere, and this was at least in part because of the way that Cuban science was positioned outside of the dominant do-ability paradigm of cancer research in the West. … The Cuban work was thus at the very margins of the already marginal (Reid-Henry 2010, 100).

These choices, moreover, revealed the adoption by Cuban physicians and biotechnologists of a systemic view of the human body and its health, according to which the "pathological" was considered and treated as a part of the "normal" functioning of the immunologic system, to be in some sense redirected, instead of as an anomaly to be eradicated.

6.5 More Recent Achievements

Just to recall some data, according to Cárdenas (2010) in 2006 the Western Havana Bio-Cluster by itself was employing 12,000 workers and more than 7000 scientists and engineers (Lage 2006) with 91 products/projects focused on health (33 vaccines against infectious diseases, 33 oncological products, 18 cardiovascular products, 7 products for other diseases), 200 patents registered in Cuba, and 1800 international patent applications, and technological transfer with Brazil, China, Vietnam, and other countries.

Recently, at the end of 2012, the Biotechnology and Pharmaceutical Industry Group 'BioCubaFarma' was created with the aim of promoting potential businesses dedicated to supplying medical services, the greatest export potential of the Cuban economy.

Currently, BioCubaFarma is composed of 38 Biotech and Pharmaceutical Enterprises (including the Genetic Engineering and Biotechnology centre, the Molecular Immunology centre, the Cuban Neuroscience centre, Immunoassay centre, the Finlay Institute, etc.). It boasts 78 production facilities, and employs almost 22,000 workers. BioCubaFarma manufactures more than 500 drugs, and develops a total of 91 among biotechnology products and projects: 33 for cancer, 18 for cardiovascular diseases, 33 vaccines against infectious diseases, and 7 drugs to combat diabetes and other ailments. The BioCubaFarma staff includes more than 6325 high trained university graduates, 262 doctors, 1170 masters of Science graduates, 1300 technicians, and 719 researchers.[4]

Focusing only on CIGB, it is considered a leader centre in Cuban biotechnology with about 1400 staff workers and more than 50 research and development projects linked to human and veterinary vaccines, development of therapeutic molecules, diagnosis systems, genomic, proteomic and bioinformatics, etc. According to its 'Business Portfolio 2014–2015',[5] CGIB owns 70 inventions worldwide and is working, together with Heber Biotec S.A. (the commercial arm of CIGB, that closes the cycle from research to commercialization of CIGB results), on the introduction of novel products into the most regulated markets such as US, Europe, Canada and Japan, promoting early stage partnerships for joint development, and sharing commercial opportunities with development and sharing commercial opportunities with partners.

CIGB scientific and production activities are developed in close collaboration with other institutions of BioCubaFarma, and the National Health and Agriculture Ministries and, in parallel, with the global oriented commercial strategy of the BioCubaFarma High Level Management Entrepreneurial Organization (Castillo et al. 2013).

The number of new drug applications in the CIGB remained stagnant (but did not decrease) between 1993 and 1997, and after 1998 started to grow again at an impressive rate, tripling in 2005 (López Mola et al. 2006). From 1991 to 2005 the incidence of acute hepatitis B in Cuba fell practically to zero, and disappeared altogether in children under 15 (see López Mola et al. 2006).

As a report of the World Bank states,

> [...] the growth of the local pharmaceutical industry, which by the mid-1990s was bringing Cuba some 100 million dollars a year in export earnings, has not only covered domestic demand for medicines, but has also led to the development of products that compete on the international market. Cuba is the only country in the world, for example, that has come up with an effective vaccine against meningitis B (Kaplan and Laing 2005).

---

[4]http://oncubamagazine.com/economy-business/biocubafarma-unite-and-conquer/. Last access March 15, 2016.

[5]http://www.cigb.edu.cu/extranet/portafolio/Business_Portfolio2014-2015.pdf. Last access March 15, 2016.

This success is widely acknowledged.

In the wake of the Soviet collapse, Cuba got so good at making knock-off drugs that a thriving industry took hold. Today the country is the largest medicine exporter in Latin America and has more than 50 nations on its client list. Cuban meds cost far less than their first-world counterparts, and Fidel Castro's government has helped China, Malaysia, India, and Iran set up their own factories: "south-to-south technology transfer" (Starr 2012).

About the relationship between Cuba and the United States, at least in the past, it could be of some interest to recall some controversial reciprocal allegation of biological warfare. In 1996 Cuba reported that a US State Department aircraft designed to eradicate narcotics crops, and authorized to fly across Cuban airspace to Colombia, sprayed an unidentified substance over a Cuban area where a *Thrips palmi*[6] epidemic broke out a few weeks later. This was really not the only occasion in which Cuba raised allegations of this kind, but in this case, with Russia making the request on behalf of Cuba, it was submitted to the countries that have signed the 1975 Biological Weapons Convention.[7] In fact, the Convention had no power to promote further investigations or impose sanctions; indeed, the controversy served to make it clear that the Convention lacks a legally binding verification regime, unlike the Nuclear Non-Proliferation Treaty and the Chemical Weapons Convention. On the part of the United States, in 2002 John R. Bolton, under-Secretary of State for arms control in the Bush Jr. Administration, publicly accused Cuba of producing small quantities of germs that can be used in biological warfare.[8] The CIA itself later rebuffed these claims.[9] Cuba's success was evidently troublesome for the United States.

---

[6]An insect pest that can cause damage to a wide range of vegetable crops.

[7]Declan Butler, U.S.–Cuba row over insects goes to weapons meeting, *Nature*, 388, 21 August 1977, http://www.readcube.com/articles/10.1038/41846 (last access March 15, 2016). U.S. Denies Spraying Biological Agent Over Cuba, *The New York Times*, May 7, 1997, http://www.nytimes.com/1997/05/07/world/us-denies-spraying-biological-agent-over-cuba.html (last access March 15, 2016).

[8]J. Miller, Washington Accuses Cuba of Germ-Warfare Research, The New York Times, May 7, 2002, http://www.nytimes.com/2002/05/07/international/americas/07WEAP.html (last access March 15, 2016). Fidel Castro, CUBA: 'Our weapons are morality, reason and ideas', May 22, 2002, https://www.greenleft.org.au/node/27449 (last access March 15, 2016). Thinktank disputes Bush administration claims of biowar development in Cuba, centre for International Policy, May 2002, http://www.afn.org/iguana/archives/2002_05/20020508.html (last access March 15, 2016).

[9]Wayne S. Smith, More Empty Charges, April 7, 2004, http://articles.sun-sentinel.com/2004-04-07/news/0404060324_1_cuba-biological-weapons-bolton-s-statement (last access March 15, 2016). L. and S. San Martin, CIA rebuffs John Bolton and Otto Reich claim of Cuba's biological warfare capabilities, Miami Herald, April 09, 2005, http://havanajournal.com/politics/entry/cia_rebuffs_john_bolton_and_otto_reich_claim_of_cuba_biological_warfare_cap/ (last access March 15, 2016). For a full account see: NTI, Country Profiles, Cuba, Biological, http://www.nti.org/country-profiles/cuba/biological/ (last access March 15, 2016).

6.6 Further Cuban Distinctive Features: South–South Cooperation, Medical Diplomacy

Besides the achievements of Cuban biotechnology, there are other, indirectly related, features that deserve at least to be mentioned in order to complete the picture of the distinctive features of the Cuban system.

From Chap. 3 on we have stressed Cuba's absolutely original strategy of international south-south cooperation, mainly (but not only) in medical diplomacy. Cuba has an exceptionally high number of health professionals serving abroad (in 2006, 28.664 in 69 countries),[10] and supports health programs in 27 countries in Latin America, the Caribbean, Africa and Asia. Among these, there is the Cuba-Venezuela "oil-for-doctors" bilateral agreement. In this context, the *Operación Milagro* (Operation Miracle) is an eye-surgery program, which is treating more than 200,000 patients from 21 countries. There is also an agreement with Japan for eye surgery and treatment of Japanese citizens in Cuba's *Centro de Retinosis Pigmentaria* (centre of Pigmentary Retinopathy). After the collapse of the apartheid regime in Angola, thanks also to the contribution of Cuba's intervention, South Africa suffered a post-apartheid brain drain (white flight), and in 1996 it began importing Cuban doctors.

---

[10]Cuba not only provides medical services to unserved and underserved communities within Venezuela (30,000 medical professionals, 600 comprehensive health clinics, 600 rehabilitation and physical therapy centres, 35 high technology diagnostic centres, 100,000 ophthalmologic surgeries, etc.), but also provides similar medical services in Bolivia on a smaller scale at Venezuela's expense…. And to contribute to the sustainability of these health programs, Cuba will train 40,000 doctors and 5000 healthcare workers in Venezuela and provide full medical scholarships to Cuban medical schools for 10,000 Venezuelan medical and nursing students. … An additional recent agreement includes the expansion of the Latin American and Caribbean region-wide ophthalmologic surgery program (Operation Miracle) to perform 600,000 eye operations over ten years. … Cuban medical teams had worked in Guyana and Nicaragua in the 1970s, but by 2005 they were implementing their Comprehensive Health Program in Belize, Bolivia, Dominica, Guatemala, Haiti, Honduras, Nicaragua, and Paraguay…. Because Cuba has been successful in developing health programs at home and has provided medical aid abroad, often under difficult circumstances, some donor countries are willing to provide financial support for Cuban medical assistance in third countries in what is called triangular cooperation. Germany has provided funding for Cuba to develop health programs in Niger and Honduras. France provided some funding to execute a health program in Haiti. Japan provided two million doses of vaccines to vaccinate 800,000 children in Haiti and U.S.$57 million to equip a hospital in Honduras where a Cuban medical brigade works. Multilateral agencies, such as the World Health Organization (WHO) and the Pan American Health Organization (PAHO) also finance medical services provided by Cuba for third countries. Both organizations provided funding for Cuba's medical education initiatives. … By 2004, there were about 1200 Cuban doctors working in African countries, including in Angola, Botswana, Cape Verde, Côte d'Ivoire, Equatorial Guinea, Gambia, Ghana, Guinea, Guinea-Bissau, Mozambique, Namibia, Seychelles, Zambia, Zimbabwe, and areas in the Sahara. … On the African continent, South Africa is the financier of some Cuban medical missions in third countries (Feinsilver 2006).

Furthermore, Cuba has not missed a single opportunity to offer and supply disaster relief assistance irrespective of whether or not Cuba had good relations with that government. This includes an offer to send over 1000 doctors as well as medical supplies to the United States in the immediate aftermath of Hurricane Katrina. Although the Bush administration chose not to accept the offer, the symbolism of this offer of help by a small, developing country that has suffered forty five years of US hostilities, including an economic embargo, is quite important (Feinsilver 2006).

It is no wonder that Cuba's concrete international engagement has been politically unwelcome not only to the United States, but often to political and corporate interests as well.

But not all are thrilled to have Cuban doctors in town. In particular, local medical associations and individual doctors have harshly criticized the Cuban presence because of their competition for jobs, their different manner of working and treating patients, and because of the perquisites they receive (principally, free room and board). In some cases, such as in Bolivia and Venezuela, these medical associations have gone on strike to protest the Cuban presence…. Despite protests (and strikes), numerous press and other reports from different countries extol the benefits to the patients, many of whom had never seen a doctor before, particularly living and working in their own neighbourhood (Feinsilver 2006).

It is not by chance that the Venezuelan counter-revolutionary movements single out for criticism not only Cuban doctors, but also health structures and equipment set up by Cuba to assist citizens.

Rather than a fifth column promoting socialist ideology, these doctors provide a serious threat to the status quo by their example of serving the poor in areas in which no local doctor would work,… changing the nature of doctor-patient relations. As a result, they have forced the re-examination of societal values and the structure and functioning of the health systems and the medical profession within the countries to which they were sent and where they continue to practice. This is the current Cuban threat … (Feinsilver 2006).

As we have warned in the Introduction of this book, the new neo-liberal political majority which won the December 2016 elections in Venezuela (such as the new Argentinian President) have explicitly expressed the will to put an end to all kinds of cooperation with, and support from Cuba. While we are writing, it is too early to foresee what will happen, but there is the possibility that in the future the *Operación Milagro* with Venezuela stops.

The political implications of this medical diplomacy are even more complex, both for Cuba's domestic policy and its international relations.

The temporary export of Cuban doctors also provides a safety valve for disgruntled medical professionals who earn much less at home than less skilled workers in the tourism sector. Their earning opportunities abroad are significant both within the confines of medical diplomacy and even more so, beyond it. This has led to a number of defections, allegedly around six hundred, although some say this figure is too high. This figure could grow if Cuban-American activist groups carry out their threats to assist these doctors serving in foreign lands if they defect. Should this number increase dramatically in this period of political change, the Cuban government may decide that the cost is too great to bear (Feinsilver 2006).

But Cuba's medical diplomacy is not limited to its interventions abroad. Another very remarkable initiative was the creation in Cuba in 1999 of the Latin American School of Medicine (ELAM), operated by the Cuban government. This was Cuba's answer to hurricanes George and Mitch, which in 1988 deeply affected the economies of the Caribbean countries and claimed thousands of victims, after sending brigades of health personal to the most affected areas.

The ELAM has been described as possibly the largest medical school in the world for enrolment. Offering free tuition, accommodation and board, it enrols thousands of students from 122 countries of Latin America and the Caribbean, Africa and Asia[11] (the school also accepts minority students from the United States, and some dozens are reported as enrolled). In exchange for full scholarships, these students must be willing to return to their countries and practice medicine in poor communities for at least five years. In this way Cuba helps turn the "brain drain" into a "brain gain."

According to the authoritative UNESCO Science report launched in November 2015[12]

Cuba is one of the most popular student destinations within Latin America; the UNESCO Institute for Statistics estimates that there are around 17 000 students from other Latin American countries living in Cuba, compared to 5 000 in Brazil and around 2 000 in each of Argentina and Chile.....While eleven Latin American countries devote more than 1 % of GDP to higher education, the 4.47 % of GDP spent on higher education by Cuba, represents the highest in the region.....The Cuban economy grew by 1.3 % in 2014 and is expected to expand by 4 % in 2015. In 2014–2015, 11 priority sectors for attracting foreign capital were identified, including agrifood; general industry; renewable energy; tourism; oil and mining; construction; and the pharmaceutical and biotechnology industry.[13] With the normalization of relations with the US in 2015, Cuba is in the process of establishing a more attractive legal regime offering substantial fiscal incentives and guarantees for investors. .... In 2014, the government created the Financial Fund for Science and Innovation (FONCI) to enhance the socio-economic and environmental impact of science by boosting business innovation. This is a major breakthrough for Cuba, considering that, up until now, the bulk of R&D funding has come from the public purse.

---

[11] See for instance http://www.nnoc.info/latin-american-school-of-medicine/ (last access March 15, 2016) and Don Fitz, The Latin American School of Medicine today: ELAM, *Monthly Review*, 62 (10), 2011.

[12] http://www.unesco.org/new/en/natural-sciences/science-technology/prospective-studies/unesco-science-report/unesco-science-report-2015/ Last access March 15, 2016.

[13] ECLAC (Economic Commission for Latin America and the Caribbean), *Economic Survey of Latin America and the Caribbean. Challenges in boosting the investment cycle to reinvigorate growth.* Santiago.2015. http://www.cepal.org/en/publications/38715-economic-survey-latin-america-and-caribbean-2015-challenges-boosting-investment Last access March 15, 2016.

6.7 Cuba's Remarkable and Enduring Achievements

As we have discussed and substantiated with specific and authoritative references, the achievements of Cuban science are unquestionable, and widely acknowledged worldwide. In this chapter we have discussed how the collapse of the Soviet Union and the Socialist market strained the country but, contrary to the majority of the forecasts, Cuba overcame this terrible challenge. In particular the scientific system that had been created in the decades 1960s–1980s withstood the repercussions of terrible blow, resorting to the multiple and open relations established in the previous decades outside the Socialist countries. As we have discussed in details, the field of biotechnology was created in the 1980s independently from the Soviet Union, that was quite backward in the fields of modern genetics and molecular biology.

The big emergence in which Cuba found itself in the early 1990s was resolutely tackled confirming, and even reinforcing, despite the deep economic difficulties, the same strategy that Cuba had originally chosen at the very beginnings of the revolution, to overcome the condition of subalternity and achieve economic development: in Gramsci's words, "conquering 'ideologically' the traditional intellectuals" (Sect. 1.5). As we have already quoted,

> Faced with economic calamity, Castro did something remarkable: he poured hundreds of millions of dollars into pharmaceuticals (Starr 2012).

Consequently, during the 1990s the sector of health biotechnology got stronger, new centres were created, coordination was strengthened, innovative processes, products and therapies were achieved, and this field became one of the main source of hard currency income for the country. These achievements confirmed the soundness of the Cuban model of biotechnology production and innovation, alternative to the capital-intensive model of multinational firms. At the same time, Cuba kept its internationalist policy of support to developing countries, of south-south cooperation, of medical diplomacy, of offer and supply of disaster relief assistance irrespective of whether or not Cuba had good relations with that government. As we have already mentioned, in 1999 the Cuban government created the Latin American School of Medicine (ELAM).

This path, however, was not an easy one. Although the Cuban market of biotechological and biomedical products was expanded to new countries, the economic crisis has exacerbated the fierce competition, and aggressive international relations grew. Despite its low capital-intensive structure, the Cuban biotechnological industry needed to be kept efficient, innovation was expensive and fresh capital was needed. An example of this new trend is the Paris-based ABIVAX biotech-company, created in 2014 by the French venture capital firm 'Truffle capital' in collaboration with CIGB, representing the first ever start-up launched on the basis of a Euro—Cuban R&D collaboration.[14] A therapeutic vaccine for hepatitis B vaccine, acquired from CIGB, could hit the market as early as 2017, while, more recently ABIVAX acquired three commercial vaccines, targeting typhoid, meningococcus and Leptospirosis, from the Finlay Institute.[15]

> Overall, it must be concluded that the results from developing biotechnology in Cuba have been rewarding. A new industry has been created in a developing country, which is supplying cutting-edge technology products to its people, and is generating significant profits from overseas sales in spite of severe financial constraints. Certainly, there is no evidence showing that a similar scientific, social, and economic phenomenon has taken place in any other country. Similarly, the possibility of a continuous development of this sector of the Cuban economy suggests a promising future for the solution of ongoing national problems (López Mola et al. 2006).

However, if the Sisyphus metaphor to describe the recurrent difficulties in creating endogenous research and innovation perfectly fits with developing countries (Sagasti 2004), it also applies in the case of Cuba. In fact, the same Cuban government, through the Cuban Academy of Science, launched in 2012–2013 an in-depth self-critical inquiry on the state, efficiency and problems of its own system of science, technology and innovation. This in-depth inquiry involved more than one hundred members of the Academy, produced a report, whose main conclusions were the following ones (Avedaño 2014). A tendency is observed towards a reduction of the scientific personnel created by the Revolution, with critical situations in some areas, while the formation of doctors is considered inadequate and belated, especially in areas with the most direct economic and social impact. Financing is decreasing, while the material conditions for research are growing worse, especially in the universities. Low productivity of publications and patents was denounced, scarce economic impact of science in most economic sectors, and a scarce transfer of scientific research in the technological components of exports are underscored. The present structure of the Cuban scientific system had been shaped in the past, and if on the one hand it had allowed the country to overcome the critical problems of the 1990s, on the other hand these problems created wounds and negative consequences. At present the system no longer seems adequate in the context of a profoundly changed reality. The report declared that a new phase of growth is needed, carefully planned to meet the new needs with rational criteria that can define quantifiable indicators, and reshape the strategy of financial support of the system, balancing state sources with those of corporate origin, which have different functions. Even the educational system, according to the Cuban specialists, needed to be revamped and modernized, providing more stimuli from elementary education up, and the relaunching, modernization, and greater integration of higher education. Moreover, the report suggested concrete measures aimed at preserving

---

[14]http://www.abivax.com/en/com-abivax-title-medias/news-events/press-releases/23-creation-of-abivax-a-leader-in-therapeutic-vaccines-and-the-first-french-company-to-sign-an-exclusive-partnering-agreement-with-cuba-in-healthcare.html. Last access March 15, 2016.

[15]http://www.fiercevaccines.com/story/abivax-eyes-49m-ipo-advance-cuban-made-hep-b-vaccine/ 2015-06-11 Last access March 15, 2016.

and increasing the efficiency and impact of Cuban science as concerns scientific management and human and financial resources.

A process of revision was undertaken along these lines, in order to formulate political proposals aimed at reorganizing the Cuban System of Science Technology and Innovation. In the near future profound changes in Cuban science could have been expected.

Just at this stage Obama's thaw towards Cuba unexpectedly supervened, and completely changed the situation. As we have already discussed in the Introduction of this book, nobody can foresee what the future will reserve. But it is certain that nothing will be as before. For that reason we have stopped our reconstruction to the end of 2014. In any case Cuban achievement of the construction of a modern and efficient scientific system, including such competitive fields as healthcare and biotechnology undertaking, is a story that is worth sharing.

References

Avedaño B (2014) Panorama científico Cubano. Escuchar, privilegio de la sabiduría. *Bohemia*. 29 Sept http://www.bohemia.cu/2014/09/29/encuba/ciencia.html. Last access 15 Dec 2014

Baracca A, Renn J, Wendt H (eds) (2014) The history of physics in Cuba. Springer, Berlin

Cabal Mirabal CA (2014) Magnetic resonance project 35-26-7: a Cuban case of engineering physics and biophysics (Baracca A, Renn J, Wendt H, eds), pp 315–322

Cárdenas A (2009) The Cuban biotechnology industry: innovation and universal health care. https://www.open.ac.uk/ikd/sites/www.open.ac.uk.ikd/files/files/events/innovation-and-inequality/andres-cardenas_paper.pdf. Last access 15 March 2016

Cárdenas A (2010) The Cuban biotechnology: innovation and universal health care. In: Innovation and inequality workshop, 15–16 May 2010, Pisa, Italy, http://www.open.ac.uk/ikd/sites/www.open.ac.uk.ikd/files/files/events/innovation-and-inequality/andres-cardenas_presentation.pdf. Last access 15 March 2016

Castillo A, Caballero A, Triana J (2013) Economic-financial management modeling for biotechnology enterprises in Cuba. Biotecnología Aplicada, 30(4):290–297. ISSN 1027-2852

Elderhost M (1994) Will Cuba's biotechnology capacity survive the socio-economic crisis? Biotecnol Dev Monitor 20 (Sept), 11-13/22

FAO, Food and Agriculture Organisation (2003) FAO Statistical Database (Food and Agriculture Organization of the United Nations) Rome. http://apps.fao.org

Feinsilver JM (1993) Can biotechnology save the revolution?. NACLA Rep Am 21(5):7–10

Feinsilver JM (1995) Cuban biotechnology: the strategic success and commercial limits of a first world approach to development. In: Peritore NP, Galve-Peritore AK (eds) Biotechnology in Latin America: politics, impacts and risk. Scholarly resources. Wilgminton DE, pp 97–126

Feinsilver JM (2006) La Diplomacia Medica Cubana: Cuando La Izquierda Lo Ha Hecho Bien. Foreign Affairs 6(4):81–94 (English transl: Cuban medical diplomacy: when the left has got it right). http://www.coha.org/cuban-medical-diplomacy-when-the-left-has-got-it-right/. Last access 15 March 2016

Giles J (2005) Cuban science: ¿vive la revolution? Nature 436(21 July):322–324

Kaplan W, Laing R (2005) Local production of pharmaceuticals: industry policy and access to medicines. In: Health, nutrition and population discussion paper. The World Bank, 16 Jan

Kirkpatrick AF (1996) Role of the USA in the shortage of food and medicine in Cuba. The Lancet 348:1489–1491

Lage A (2006) The knowledge economy and socialism: is there an opportunity for development? Rev Cuba Socialista. 41: 25–43.

Lage A (2009) Transforming cancer indicators begs bold new strategies from biotechnology. MEDICC Rev 11(3):8–12

Lage A (2014) Immunotherapy and complexity: overcoming barriers to control of advanced cancer. MEDICC Rev 16(3–4):65–72

López Mola E, Silva R, Acevedo B, Buxadó JA, Aguilera A, Herrera L (2006) Biotechnology in Cuba: 20 years of scientific, social and economic progress. J Commercial Biotechnol 13:1–11

ONE (1996) (Oficina Nacional de Estadísticas). Anuario Estadístico de Cuba 1996. Cuba Edición Oficina Nacional de Estadísticas, Havana, Cuba, 1998

Peña Y, Perera A, Batista JF (2014) Immunoscintigraphy and radioimmunotherapy in Cuba: experiences with labeled monoclonal antibodies for cancer diagnosis and treatment (1993–2013). MEDICC Rev 16(3–4):55–60

Plahte J, Reid-Henry S (2013) Immunity to TRIPS? Vaccine production and the biotechnology industry in Cuba. In: Löfgren H, Williams OD (eds) The new political economy of pharmaceuticals: production, innovation and TRIPS. Pelgrave Macmillan, pp 70–90

Reid-Henry S (2010) The Cuban cure: reason and resistance in global science. University of Chicago Press, Chicago

Sagasti F (2004) Knowledge and innovation for development. the sisyphus challenge of the 21st century. Edward Elgar Publishing, Cheltenham (UK)

Sanchelima J (2002) Selected aspects of Cuba's intellectual property laws. Cuba in transition: volume 13. Association for the study of the Cuban economy (ASCE) 213–219. http://www.ascecuba.org/publications/proceedings/volume12/pdfs/sanchelima.pdf. Last access 15 Dec 2015

Starr D (2012) The Cuban biotech revolution. http://www.wired.com/wired/archive/12.12/cuba_pr.html. Last access 15 March 2016

Chapter 7
Comparative Considerations and Conclusions

Compared with most other countries, the business of Cuban biotech is exceptional for one simple reason: it has been an exclusively state-sponsored enterprise. Indeed, Cuba has a long and distinguished history in biotech due to Fidel Castro's commitment to developing science in the country. [Buckley et al. 2006]

Abstract The noteworthy success of a small embargoed island in scientific development, and in particular in a typically US-dominated and capital-intensive sector like biotechnology, has attracted considerable interest and discussion among the analysts and specialists, since it shows features that are unique in the panorama of developing countries. Cuba's achievements in science and technology seem an exception with respect to what usually happens in other underdeveloped countries, excluded probably the biggest and richest ones. Even more exceptional is the development of biotechnology in Cuba. Some concepts are summarized, inspired form the most competent specialists in the field.

Keywords Biotechnology industry · Integration *versus* competition · Public research institutions · Full-cycle research-production · *Empresa Estatal Socialista de Alta Tecnologia* · Biotechnology in third worlds countries · Brasil · South Korea

7.1 The Intriguing Issue of Cuba's Scientific Achievement: Knowledge-Based Economy and State High Technology Company

Generally speaking, biotechnology is the quintessential capital-intensive product of advanced financial capitalism, it introduced an imperial relationship with nature which has opened the door to the proprietary ownership of living matter. The material interests that underlie it, shape the very approach of biotechnology. Yet a small country like Cuba, with limited resources, has developed a successful, cost-effective

© The Author(s) 2016   93
A. Baracca and R. Franconi, *Subalternity vs. Hegemony, Cuba's Outstanding Achievements in Science and Biotechnology, 1959–2014*, SpringerBriefs in History of Science and Technology, DOI 10.1007/978-3-319-40609-1_7

and efficient alternative to this world dominant approach. Almost two decades ago, in the most dramatic economic situation imaginable in Cuba, a specialist remarked:

> one must ask why and how a small developing nation like Cuba could even contemplate the use of biotechnology as part of a national economic survival strategy. Even among Western industrialized countries, only Japan made biotechnology part of its national development strategy. Moreover, few biotechnology companies in the United States are successful, and all are seeking alliances with transnational pharmaceutical companies in order to gain access to capital and marketing networks (Feinsilver 1993b).

The reasons of the Cuban success in this field have attracted considerable interest and discussions among the specialists in the field (Feinsilver 1993a, 1995; Elderhost 1994; Kaiser 1998; Thorsteinsdóttir et al. 2004b, c; Giles 2005; Buckley et al. 2006; López Mola et al. 2006, 2007; Evenson 2007; Editorial 2009; Lantigua Cruz and González Lucas 2009; Cárdenas 2009; Reid-Henry 2010; Scheye 2010; Starr 2012). But before trying to summarize the arguments brought in these studies, we would like to start with an absolutely general consideration.

Must we really wonder of the swift progress of science in Cuba since the 1960s, and in particular of the almost sudden development of biotechnology at an international standard? How was it possible? Was it a unique case? Cubans are not extra-terrestrial creatures, gifted with superior intelligence or skills. They are on the average absolutely normal persons. In our opinion and experience, some degree of inventiveness, or resourcefulness, the art of scrapping, must be acknowledged to the Cubans (essentially the same that allows to the ancient American cars to continue to circulate in Cuba, despite the lack of spear parts since almost 60 years). But this cannot be a credible explanation.

Therefore, we must change the question. Did in the situation of revolutionary Cuba exist some peculiar condition, or a mixture of conditions, which provided to the Cubans particular motivations or stimuli that stirred their creativity? From that standpoint, various arguments can be proposed.

In the first place, the success of the Cuban revolution put the country in complete contrast with the most powerful imperial power. The Cubans have a dose of pride. Not only the survival, but even the success of the revolution became in some way a goal that, galvanized by Fidel Castro and the revolutionary leadership, was picked out by all the Cubans (obviously, those who did not leave the country) like a challenge, or a bet, in which the whole population put all its willingness, talent and fantasy. It seems at least plausible that a sort of collective will arose, which multiplied forces and opportunities. In particular, the speeches of Fidel strongly pushed in this direction, as well as the (however, or precisely because, strongly idealistic) "Che" Guevara's voluntary work and moral stimuli, establishing an effective *hegemony* (in Gramsci's words, Sect. 1.5, "conquering 'ideologically' the traditional intellectuals").

In this context, in particular the Cuban scientific community was loaded with social responsibilities and goals that presumably strongly stimulated their will. In a sense, the usual ideology of the progressive role of science, which is generally assumed in an abstract sense by the scientific community, developed concrete tasks and commitments.

In general terms, Cuban science developed a peculiar model of scientific organization and structure. It deeply differed, in our opinion, not only from the privatistic
organization of capitalistic countries, but in several aspects also from the centralized
direction of the Soviet organization. With respect to the first one, the social tasks and
the public needs were prioritized with respect to the individual careers and interests,
and the hierarchies of power. With respect to the second one, there was in Cuba at the
same time a collective and effective participation of the whole scientific community
(even, in the initial times, of the student component, which was training to the
scientific profession) to the basic decisions, and an enlightened and constructive
planning from the political establishment, with the result that the social tasks
prevailed. In particular, the career logic of scientific promotion and the privileges of
the scientific elite were practically absent in Cuba (even though they acted as a
mermaid for several scientists who left the country). Who has collaborated with
Cuban scientists and scientific organizations should have been struck by the absence
of competitiveness and rivalries, and from the highly collaborative spirit.

We shall try to summarize the main arguments discussed in the specialistic
analyses we have cited.

Some of the factors that have made possible these accomplishments are: the
availability of qualified human resources

a country of men and women of science,

a product pipeline already supplying the domestic health system and a growing
export capacity, the design of facilities as integrated research-production organizations able to close the loop from research to the economic return, state guidance,
social ownership, export orientation, and the comprehensive integration of the
Cuban biotechnology multi-institutional system (Lage 2000). In a recent analysis
(Lage 2013) the results of the Cuban biotechnology, in relation to their medical and
scientific benefits as well as the features of the high-level state scientific organization (*Empresa Estatal Socialista de Alta Tecnologia*, Socialist State High
Technology Company), are discussed. Indeed, it cannot be denied that in Cuba
biotechnology essentially represents a peculiar socio-economic experience of
building connections between science and economy. The most influential among
Cuban scientists, Augustín Lage, has elaborated in an original way the concept of
"knowledge-based economy" (*economía del conocimiento*), meaning the direct
transformation of knowledge into economic value, as a sort of substitute for economic capital (Lage 2006). As the concept is developed in a recent paper:

> Valorization of knowledge generated by the fast advance of science and its marketing has
> led to the so-called 'knowledge-based economy' and as a consequence high-technology
> (HiTech) companies have emerged. Those based in Biotechnology have some specific
> features compared to other HiTech sectors. ... [Our] model is based on some concepts and
> proposals, and it addresses key elements, such as: insurance of adequate funding levels for
> R&D activities and technology replacement, flexible import and export management, the
> ability to get into very competitive markets and preservation of a highly qualified work-
> force. It also demonstrates the feasibility to establish this kind of enterprises in the context
> and regulations already existing in a non-capitalist environment, in the middle of the update
> of Cuban economy (Castillo et al. 2013).

7.2   Peculiar Features of Cuban Biotechnology Industry

Coming now to a systematic analysis of the peculiar features of Cuban Biotechnology industry, as they emerge from the specialistic literature, the following aspects seem relevant:

- The Cuban government had an unusual level of commitment to scientific and technological development, and a long-term vision to support health biotechnology, despite difficult economic conditions. Cuba took a first-world approach to the rapid generation and application of science and technology for economic development (Feinsilver 1995, 98).
- The Cuban government has placed social policy at the centre of its development policy. As a consequence, social needs have been guiding criteria for the choices and actions of the Cuban government.
- The Cuban government considered the development of universal education, free higher education, strong scientific training, and scientific research a pre-condition.
- Access to an educated workforce and a well-functioning public health system contributed to innovation:

  The educational level of Cuba's labour force at the end of the 1990s was almost equal to that of the Organization for Economic Cooperation and Development (OECD) countries (Thorsteinsdóttir et al. 2004b, 21).

- Public research institutions form the backbone of health biotechnology:

  some university institutions have made impressive contributions to health biotechnology. For example, researchers from the Faculty of Chemistry at the University of Havana made a leading contribution in the development of the synthetic *H. influenzae type b* vaccine[1] (Thorsteinsdóttir et al. 2004b, 21).

---

[1]*Haemophilus influenza*e type B, or Hib, is a bacterium estimated to be responsible for some three million serious illnesses and over 350,000 deaths per year, chiefly through meningitis and pneumonia. Almost all victims are children under the age of five, with those between four and 18 months of age especially vulnerable. Hib meningitis is a more serious problem in developing countries, with mortality rates several times higher than seen in developed countries; it leaves 15 to 35 % of survivors with permanent disabilities such as mental retardation or deafness. However, Hib is preventable—highly effective vaccines have been available since the early 1990s. Yet hundreds of thousands of children die year after year from Hib disease. One major reason is that the Hib vaccine is significantly more expensive than other childhood vaccines; for a low and middle income country the Hib vaccine costs roughly seven times the cost of vaccines against measles, polio, tuberculosis, diphtheria, tetanus, and pertussis combined (about $7 USD versus $1 USD). In Cuba, national research and policy organizations have joined forces to implement an integrated strategy governing vaccines from the development stage to the distribution stage. This strategy brings together institutions involved in every life stage of a vaccine (including government ministries, clinical research organizations, support institutions and manufacturing facilities). [...] Thanks to the development of capacities and facilities to internalize the entire supply chain of vaccines, Cuba has been able to develop various vaccines and antibiotics at low cost while ensuring distribution of these life-saving advances throughout the country. In 1999, the first

- Many public research institutes house diverse health biotechnology activities, including research, development and production.
- Knowledge sharing and flow among and within research institutions have been an important stimulus for innovation (integration *vs* competition).
- Cuba's health biotechnology research system has strong links with the country's public health system, which is not only the recipient of innovation but also a contributor. This has encouraged collaboration between basic and clinical researchers, and has promoted the adoption of cost-effective treatment options.

Some features have been enhanced, or introduced in the past two decades, in response to the crisis caused by the downfall of the Soviet Union, as for instance:

- Integration and collaboration within the institutes of the Scientific Pole of Havana (or Western Havana Bio-Cluster) have been reinforced.
- Greater emphasis has been placed on innovation from within Cuba (Thorsteinsdóttir et al. 2004b, 19).

Many of the institutes have developed the whole cycle or the closed cycle, from research, through development, production, quality control and commercialization of the end products. Some institutions have even acquired a commercial arm, often in the form of an associated company (Thorsteinsdóttir et al. 2004b, 21).

- Cubans have been active in licensing and setting up strategic alliances and joint ventures based on Cuban biotechnology with companies around the world:

In addition to collaborative ventures with Canada and Great Britain, Cuba presently licenses its biotechnology and has joint projects in a growing number of developing countries around the world, including Algeria, Brazil, Canada, China, India, Malaysia, Mexico, South Africa, Tunisia, and Venezuela. In addition, China has become a major participant in Cuban biotech projects; three joint ventures have been formed in China using Cuban technology (Evenson 2007).

(Footnote 1 continued)

commercial vaccine containing a synthetic carbohydrate antigen was developed in Cuba against Hib. This vaccine, Quimi-Hib (Heber Biotech), exhibits several advantages over naturally-derived vaccines, such as lower production costs compared with conventional vaccines, and higher quality control standards compared with naturally-derived agents. In clinical trials, also conducted in Cuba, researchers found that the HiB vaccine provides protection to nearly 100 % of immunized infants after primary vaccination and a second booster dose. Additionally, clinical trials showed the vaccine to be very safe. This lower-cost alternative provides access to the Hib vaccine for those who otherwise would not have been able to afford it.» (Pan American Health Organization, and World Health Organization, Vaccine Research and Development in Cuba, http://www.paho.org/hq/ index.php?option=com_content&view=article&id=3114%3A2010-vaccine-research-development-cuba&catid=6601%3Akbr-case-studies&Itemid=40275&lang=en. Last access March 15, 2016.

As a result,

- Researchers in Cuba have filed about 500 patent applications in the health biotechnology sector based on more than 200 inventions (according to an analysis of the European Patent Office's (Munich, Germany) database: the European Network of Patent Databases, May 2003, http://www.european-patent-office.org/). These have been filed in several countries throughout the world, including the United States, Europe, Brazil, India, China and South Korea. Cuba exports biotechnology products to more than 50 countries, mainly in Latin America, Eastern Europe and Asia (Thorsteinsdóttir et al. 2004b, 19).

Cuba is now a location of choice, with major global pharmaceutical companies opening offices in Cuba (Scheye 2010).

7.3 Something Worth Thinking Seriously About: A Comparison with Other Experiences

By way of conclusion, it would be extremely interesting, and also valuable for the strategies of development and social progress of developing countries (but not only, if one takes into account the present conditions of subalternity of some nations into today's European Community) to draw some comparisons between Cuba and other Third World countries in scientific development and its achievements.

A general consideration drawn in a study of the late 1980s, when Cuban biotechnology was booming, still sounds relevant for the contrast with the strategic choices that Cuba had long since developed:

Third World countries are not pursuing scientific and technological policies leading to the development of strong biotechnological industries. Their leaders have been misled into believing that modern biotechnological industries can be built in the absence of strong, intellectually aggressive, and original scientific schools. Hence, they do not strive to reform their universities, which have weak commitments to research, and do not see the importance of having research hospitals able to generate excellent and relevant clinical investigation. These strategic gaps in scientific capability, the lack of governmental and corporate research funding, and the dependent nature of the chemical and pharmaceutical industries of the Third World make the development of competitive biotechnology a highly improbable event (Goldstein 1989).

According to Goldstein, at least at the date of his well-documented and argued study, contrary to the "biotech propaganda blitz", the tall tale of biotechnology and applied science as factors for meeting economic underdevelopment turns into its contrary, i.e. a factor of further exploitation of the economies of Third World countries, that is the defeat of the struggle to overcome subalternity. The Cuban exception could hardly be more evident.

Regarding Latin American countries, one should remark that at least some among them were not precisely "inexperienced" in the biomedical-biochemical field

(Goldstein 1989; Cueto 2006). In Argentina, Bernardo A. Houssay (1887–1971) was co-winner of the Nobel Prize for Physiology or Medicine in 1947. His school in Buenos Aires flourished with Braun-Menéndez (1903–1959), Luis Federico Leloir (1906–1987), 1970 Nobel Prize in Chemistry, and others. Another colleague of Houssay's, the Uruguayan Roberto Caldeyro-Barcia (1921–1996), pioneered the field of maternal-fetal medicine in Montevideo. Daniel Vergara Lope (1865–1938) in Mexico and Carlos Monge Medrano (1884–1970) in Peru made important contributions to high altitude physiology. In Brazil Maurício da Rocha e Silva (1910–1983) brought outstanding contributions to pharmacology, and was the chief architect of the development of this discipline. Guillermo Whittembury in Venezuela contributed to modern kidney physiology. However, in several Latin American countries the development of these disciplines, and of science in general, suffered deeply from the existence of military and repressive regimes. An exemplary case was the Argentinian pioneer in antibody research, and future 1984 Nobel Prize winner in Physiology and Medicine, César Milstein (1927–2002), who was exiled from the country in 1963 with his collaborators while he was trying to create the first group in molecular biology in the continent. He subsequently took British citizenship.

As a matter of fact, in the last few decades several developing countries have decidedly entered the business of biotechnology, among them the major Latin American ones, with different degrees of success.

Specific comparisons have been made between the development of biotechnology in Cuba and in some countries, other than the most industrialized ones, typically classified as

lower income countries or developing countries, each at a different stage of economic development when compared with industrially advanced nations» (Thorsteinsdóttir et al. 2004a, c; also Peritore and Galve-Peritore 1995).

However, we are aware that, in order that these comparisons make sense, they should consider, in the first place, the great difference between Cuba's "dimension"—economy, resources, population, and so on—and the majority of the other countries taken into account. Moreover, Cuba has a very peculiar geostrategical and political situation, not to mention the unique economic constraints due to the US embargo.

In this perspective, even more exceptional is the fact that, as we did already remark, Cuba kept increasing investments in health and medicine in the 1980s and the early 1990s, while the politics of economic austerity and financial constraints was predominant in the continent, and has confirmed this support as a strategic choice even in the extremely difficult conditions following the downfall of Soviet aid and trade.

Regarding specific features, in contrast with the level of integration of the Cuban biomedical system, in the other countries,

... lack of collaboration and linkages among health biotechnology institutions restrained innovation efforts. In China, lack of collaboration prevented its scientists from being the first in the world to sequence the severe acquired respiratory syndrome (SARS) virus. Lack of linkages, especially between universities and industry, has also slowed innovation efforts in Brazil and Egypt (Thorsteinsdóttir et al. 2004c, 50–51).

In Brazil, moreover, along with some public research institutions, universities are the main actors in health biotechnology.

Knowledge flow to and from Brazilian universities and public research institutes is however limited, as they are not well connected to enterprises (Ferrer et al. 2004).

Governmental policies are deficient.

Brazilians get lost between basic research and its transformation into technology, between academic life and the manufacturing system (Thorsteinsdóttir et al. 2005, 102).

University professors are often skeptical about close associations with companies. For their part, private sector firms lack linkages.

Particularly interesting seems a comparison between Cuba and South Korea, a country that was created after the Korea war (1950–1953), almost at the same time of the new revolutionary Cuba. Even South Korea has explicitly aimed to technological and scientific growth for its economic development, especially applied sciences, with a strong support from the United States, which conceived the country as a bastion against Communism (something symmetrical to the conception of the Soviet Union with respect to Cuba). South Korea has especially developed electronics and nuclear technology, reaching fore-runner levels, but also a strong healthcare biotechnology sector was promoted. Along with recent specific studies (Park and Leydesdorff 2010; Kwon et al. 2012), the connections between university, industry and the government in South Korea apparently reveal serious deficiencies. In fact, the inter-institutional

collaboration pattern, as measured by co-authorship relations in the *Science Citation Index*, noticeably increased, with some variation, from the mid-1970s to the mid-1990s. However, inter-institutional collaboration in the first decade of the 21st century was negatively influenced by the new national science and technology (S&T) research policies that evaluated domestic scientists and research groups based on their international publication numbers rather than on the level of cooperation among academic, private and public domains. The results reveal that Korea has failed to boost its national research capacity by neglecting the network effects of science, technology, and industry (Park and Leydesdorff 2010).

South Korea seems already a difference with respect to Cuba. A closer comparison of the fields of biotechnology (Wong et al. 2004) shows that in South Korea the healthcare biotechnology sector was promoted as a future source of economic wealth. This inflated political and investor expectations, with insufficient awareness of the high-risk nature of the field, and consequent danger that many enterprises fail in the process. Successful reverse engineering, combined with a comparatively inexpensive workforce, enabled South Korean companies to produce quality goods at a lower cost. In contrast, R&D in academia and industry did not place enough emphasis on innovation. Despite the positive indicators surrounding prospects of the sector, a single major technological and commercial breakthrough that will place South Korean biotechnology in the same league as that of the United States or the United Kingdom has not yet appeared. Despite government investments in the sector, investors seem sceptical, especially after the venture mini-bubble of the late 1990s burst.

South Korea must evolve from the industrial learning paradigm to a new technology creation paradigm. For academics and policy makers, this sort of transition makes intuitive sense. For South Korean scientists, investors, entrepreneurs and the public, however, this paradigm shift is not simply an academic problem, nor easily manipulated through top-down policy instruments. Rather, at its most basic level, the move toward technological creativity requires an attitudinal shift. It cuts to the core of the post-war South Korean mind-set. Indeed, this may prove to be South Korea's biggest challenge in making it in biotechnology (Wong et al. 2004, 46).

A general remark about the Third World is that there,

biotechnology … is a bibliocentric creed, in which the practitioners limit themselves exclusively to relearning technologies invented by others. Universities do not train people for invention and discovery but rather to follow and repeat what has been invented elsewhere. In fact, originality and inventiveness in the Third World, are, more often than not, persecuted and punished. The social blindness regarding innovation means that the scientific uses and social exploitation of the very few relevant discoveries made in the Third World mainly occur abroad (Goldstein 1995, 42).

The contrast to scientific development in Cuba could not be more complete!

7.4 Conclusions

In this book we have integrated our past and present experiences of active collaboration with Cuban scientists, and of research on Cuban science, with the most influential analyses of Cuban biotechnology accumulated in recent decades by the specialists in the field. We hope therefore to have reconstructed and analyzed in a convincing and complete way the uniqueness of Cuba's endeavour to face the high technology challenge, an endeavour based on an alternative concept and optimization of the human resources of Cuban society. Though at times our personal feelings may have shown through in the words we use, this does not invalidate the objectiveness of our main conclusions, which are not a matter of words but of facts, that we feel to have exhaustively quoted. Whatever may be one's personal opinion on Cuba, we strongly feel that the relevance of the country's achievements deserves acknowledgement, as well as the original features of its experience.

Cuba's endeavour to develop in a surprisingly short time an advanced, multi-disciplinary and polycentric scientific system has no equal in developing countries of comparable size. The achievement of an autonomous level, on equal footing in collaboration and interchange with scientists and institutions in the most advanced countries, was confirmed by the resilience of the Cuban system under the tremendous shock of the collapse of the Soviet Union and the Socialist block. This event repeated the challenge of overcoming the risk of falling back into a situation of subalternity. Once again, Cuba had to rely on its own resources, in the most difficult situation of isolation and an even more total embargo. Once more the challenge was overcome by revamping the scientific system, obviously selecting the sectors and the aims to privilege. In particular, biotechnology was confirmed as one of the backbones of Cuba's economic system.

At present Cuba faces a completely new situation. The unexpected opening by President Obama at the turn of 2014 has started a new phase, full at the same time of potential opportunities and great chances. The world political and economic situation should undergo deep transformations, besides great instabilities in the next times. Nothing will ever be as before, and no one can tell what the future has in store. For that reason we have decided to stop our reconstruction to the end of 2014. Anyhow, it seemed to us that it was a story that was worth telling.

References

Buckley J, Gatica J, Tang M, Thorsteinsdóttir H, Gupta A, Louët S, Shin MC, Wilson M (2006) Off the beaten path. Nat Biotechnol 24:309–315

Cárdenas A (2009) The Cuban biotechnology industry: innovation and universal health care. https://www.open.ac.uk/ikd/sites/www.open.ac.uk.ikd/files/files/events/innovation-and-inequality/andres-cardenas_paper.pdf. Last access 15 March 2016

Castillo A, Caballero I, Triana J (2013) Economic-financial management modeling for biotechnology enterprises in Cuba. Biotecnología Aplicada 30:290–298. ISSN 1027-2852

Cueto M (2006) Excellence in twentieth-century biochemical sciences. In: Saldaña JJ (ed) Science in Latin America. A history. University of Texas Press, Austin

Editorial (2009) Cuba's biotech boom. The United States would do well to end restrictions on collaborations with the island nation's scientists. Nature 457(January):8

Elderhost M (1994) Will Cuba's biotechnology capacity survive the socio-economic crisis? Biotecnol Dev Monitor 20(September):11–13/22

Evenson D (2007) Cuba's biotechnology revolution. MEDICC Rev 9(1):8–10

Feinsilver JM (1993a) Healing the masses. Cuban health politics at home and abroad. University of California Press, Berkely, CA

Feinsilver JM (1993b). Can biotechnology save the revolution? NACLA Rep Am 21(5):7–10

Feinsilver JM (1995) Cuban biotechnology: the strategic success and commercial limits of a first world approach to development. In: Peritore NP, Galve-Peritore AK (eds)

Ferrer M, Thorsteinsdóttir H, Quach U, Singer PA, Daar AS (2004) The scientific muscle of Brazil's health biotechnology. Nat Biotechnol 22(Supplement):8–12

Giles J (2005) Cuban science: ¿vive la revolution? Nature 436(21 July 2005):322–324

Goldstein DJ (1989) Ethical and political problems in third world biotechnology. J Agric Environ Ethics 2(1):5–36

Goldstein DJ (1995) Third world biotechnology, Latin American development, and the foreign debt problem. In: Peritore NP, Galve-Peritore AK, pp 37–56

Kaiser J (1998) Cuba's billion-dollar biotech gamble. Science 282(5394):1626–1628

Kwon K-S, Park HW, So M, Loet Leydesdorff L (2012) Has globalization strengthened South Korea's national research system? National and international dynamics of the Triple Helix of scientific co-authorship relationships in South Korea. Scientometrics 90:163–176

Lage A (2000) Las biotecnologías y la nueva economía: crear y valorizar los bienes intangibles. Biotecnología Aplicada 17:55–61

Lage A (2006) The knowledge economy and socialism: is there an opportunity for development? Rev Cuba Socialista 41:25–43

Lage A (2013) La economía del conocimiento y el socialism. La Habana: Sello Editorial Academia, ISBN 9592702861, 9789592702868

Lantigua Cruz A, González Lucas N (2009) Development of medical genetics in Cuba: thirty nine years of experience in the formation of human resources. Rev Cubana Genet Comunit [internet]. 3(2):3–23. http://bvs.sld.cu/revistas/rcgc/v3n2_3/rcgc0123010%20eng.htm. Last access 15 March 2016

López Mola E, Silva R, Acevedo B, Buxadó JA, Aguilera A, Herrera L (2006) Biotechnology in Cuba: 20 years of scientific, social and economic progress. J Commercial Biotechnol 13:1–11

López Mola E, Silva R, Acevedo B, Buxadó JA, Aguilera A, Herrera L (2007) Taking stock of Cuban biotech. Nat Biotechnol 25(11 Nov):1215–1216

Park HW, Leydesdorff L (2010) Longitudinal trends in networks of university-industry-government relations in South Korea: the role of programmatic incentives. Res Policy 39:640–649

Peritore NP, Galve-Peritore AK (eds) (1995) Biotechnology in Latin America: politics, impacts and risks. Sch Res, Wilmington, D.E.

Reid-Henry S (2010) The Cuban cure: reason and resistance in global science. University of Chicago Press, Chicago

Scheye E (2010) The global economic and financial crisis and Cuba's healthcare and biotechnology sector: prospects for survivorship and longer-term sustainability. Cuba in transition: volume 20. Twentieth annual meeting of the association for the study of the Cuban economy (ASCE) http://www.ascecuba.org/c/wp-content/uploads/2014/09/v20-scheye.pdf. Last access 15 March 2016

Starr D (2012) The Cuban biotech revolution. http://www.wired.com/wired/archive/12.12/cuba_pr.html. Last access 15 March 2016

Thorsteinsdóttir H, Quach U, Martin DK, Daar AS, Singer PA (2004a) Introduction: promoting global health through biotechnology. Nat Biotechnol 22(Supplement):3–7

Thorsteinsdóttir H, Sáenz TV, Quach U, Daar AS, Singer PA (2004b) Cuba. Innovation through synergy. Nat Biotechnol 22(Supplement):19–24

Thorsteinsdóttir H, Quach U, Daar AS, Singer PA (2004c) Conclusions: promoting biotechnology innovation in developing countries. Nat Biotechnol 22(Supplement):48–52

Thorsteinsdóttir H, Sáenz TV, Singer PA, Daar AS (2005) Different rhythms of health biotechnology development in Brazil and Cuba. J Bus Chem 2(3):99–106

Wong J, Quach U, Thorsteinsdóttir H, Singer PA, Daar AS (2004) South Korean biotechnology—a rising industrial and scientific powerhouse. Nat Biotechnol 22(Supplement):42–47

Printed in the United States
By Bookmasters